THE WARRIOR'S GARDEN

THE
WARRIOR'S
GARDEN

TOOLS FOR GUARDING YOUR MIND AGAINST BIG TECH

RICHARD RYAN

LIONCREST
PUBLISHING

THE WARRIOR'S GARDEN
Tools for Guarding Your Mind Against Big Tech
First Edition

ISBN   978-1-5445-4808-1   *Hardcover*
       978-1-5445-4806-7   *Paperback*
       978-1-5445-4807-4   *Ebook*

"There's a war going on.
The battlefield is in the mind and the prize is the soul."
–PRINCE

CONTENTS

FOREWORD

by **SAM PARR**
father, husband, human, founder of
Hampton and The Hustle (now HubSpot Media)

There's something many people don't know about the tech leaders who built Facebook, Google, YouTube—the platforms shaping our lives.

Most of them don't let their own kids use social media. They avoid giving their children smartphones entirely.

They know the truth: these tools, while transformative, are also addictive and, if left unchecked, can wreak havoc on focus, relationships, and mental health.

I've seen this play out in conversations with some of the world's most successful people—billionaires, entrepreneurs, visionaries. As the founder of Hampton, a community for high-performing business leaders, and host of the podcast *My First Million*, I've had the privilege of being part of these closed-door discussions. And one topic keeps coming up: how to regain control over technology before it takes too much from us.

Richard Ryan is someone uniquely positioned to lead this charge. His credentials speak for themselves:

- **Tech pioneer:** Richard created a YouTube app for the App Store four years before YouTube had one. Across multiple apps, he's achieved over a million downloads.

- **Media innovator:** As a media executive, he launched Rated Red with Verizon and Hearst Media, growing it to over one million organic subscribers in its first year.

- **Content creator extraordinaire:** Richard's YouTube empire includes channels like FullMag (2.7 million subscribers), with over twenty billion views across all his content.

- **Business builder**: As a co-founder of Black Rifle Coffee Company, he helped grow the brand to a publicly traded powerhouse with a $1.7 billion valuation and $396 million in revenue in 2023.

But beyond his professional accomplishments, Richard understands technology's grip on a personal level. He's lived it—felt the rush of likes and views, seen the algorithms' inner workings, and witnessed their power to distract and divide.

This book isn't about fearmongering or abandoning technology. It's about awareness. Richard doesn't just identify the problem; he offers practical tools and strategies to help you navigate a tech-dominated world with purpose and intention.

If you're reading this, you're in good hands. Richard has spent his life mastering the intersection of tech and human behavior. Now, he's ready to help you take back control.

INTRODUCTION

"Time is the only currency you spend without ever
knowing your balance. Use it wisely."
—ANONYMOUS

This is your life, and it's ending one minute at a time.

*Seven in the morning, Central Standard Time. Time to upload
that banger video! People on the West Coast will be waking up
soon, and people on the East Coast are already getting ready for
work and school.*

*I wish I could trust YouTube to publish my video, but half of the
time features are buggy or broken. Check the compression. H.264?
MP4? Good to go. Check the thumbnail. Is it gonna grab people?
Would I click on this? Maybe I should change the title. Boost the*

saturation on the skin tones and make it pop more? Nah, it's good. Okay. Publish.

Walk to the corner bakery. They have the best apple fritters. I wonder if they have Rockstar Cola. I need a Rockstar. Might have to stop at the grocery store. How many views do I have? It's at 302. Ah, probably stuck right now. Have to wait for it to update. Check the comments. Are people reacting? Is the post hitting their subscription feed?

Man, this apple fritter is awesome.

Walk back to the house. Check the video likes and comments. What's the dislike ratio? About 10 percent. That's a good ratio. I wonder why some people disliked it so quickly? Frickin' trolls. There's no way they had time to watch any of it. Maybe it's because I shot an iPhone in this one? I hate being the guy destroying a "perfectly good iPhone," but that's what gets views. My subscriber count has jumped a hundred thousand on Social Blade the last thirty days. That's good news. Looks like I won't have to reupload. Copy the video URL and paste it into emails. Write a personal note to Mike at Everyday No Days Off, MacRumors, Gizmodo, TechCrunch, and all the bloggers who have shown me love in the past. Sent! Shoot…I forgot Paul from the Awesomer. Done!

Crap, it's eight thirty. Gotta get to work!

Actually, this was my life.

Fifteen years ago I was borderline addicted to apple fritters and Rockstar Cola. I was also addicted to the rush of posting videos, monitoring responses, and watching subscriptions rise. I spent hours creating content, obsessing over thumbnails, counting likes and dislikes, and checking what others were posting.

At the time, I wouldn't have used the word *addicted*. I would have argued that I was a dedicated creator! I had to hustle if I was going to make a living off of YouTube because I was sleeping on a porch in MacArthur Park and didn't have enough money to fix my truck.

True statements, but so was the fact that I was neglecting my relationships as a result. Sure, I spent time with friends and family, but I didn't prioritize them. I didn't value them as much, and those connections atrophied as a result. Even as I say that it hurts.

This book is the wake-up call I wish I had back then.

THE TRUTH ABOUT BIG TECH

Before you slam the book shut because you think I'm telling you to get off social media, hear me out. My goal isn't to convince you to toss your phone and become a monk. Instead, I want you to stop and be honest with yourself. I want you to think about your digital usage and the ways it may be negatively impacting your life without you even realizing it. That's it. No phone burning or vow taking, though that is an option. Might even make for good content!

I can hear you through the pages: "Okay, but I'm not on my phone that much. I'm not a creator. I just check Instagram for messages a few times a day and watch YouTube videos on my lunch break. No big deal."

That might be true. Or it might not be. We'll talk about that later.

For now, we'll talk about what's happening when you're online. Let's say your friend just got back from hunting and gave you some meat for your freezer. You've been to the restaurant Dai Due before and want to be able to whip up a meal like Chef Jesse Griffiths but don't know what temperature to bring the meat to. Rather than waste your time looking through the hundreds of blogs that bury instructions among pages and pages of ads, you jump on YouTube and click on a video titled "Cooking Wild Pig."

Before you dive into this culinary masterpiece filmed in 480p, you're served up a thirty-second pre-roll advertisement on a new Amazon Prime series called *The Terminal List*, starring Chris Pratt. Epic! You loved the Joe Rogan podcast with Jack Carr, the author of the book that series is based on. But that's not why you're here.

The YouTuber finally starts talking, but not about a recipe, the meat, or the temperature. He's giving you a condensed version of his life story, along with a reminder to like, comment, and subscribe. After fifteen long minutes, he gives you the value that you sought—something that could have been explained in a couple sentences or put in the video description.

Sound familiar? Also, have you noticed that those prerolls (and mid-rolls and sponsored videos) are targeted to your interests, your location, and your behavior in very specific ways? These companies have built digital systems to identify who you are and track what you do so that they can serve you relevant ads.

Assuming you're in my key demographic, how many tampon ads do you see? I'm guessing not many, because data brokers have gathered enough information to know that you're not in the demographic that uses tampons. Ads about cold plunges, Maui Nui venison sticks, or Gorilla Mind protein powder, however—that's probably another story.

Like most people, you probably think you're driving this relationship with YouTube and other digital media.

You think you're the main character, that you're choosing what to watch and for how long. The truth is, in many ways you're just a passenger along for the ride.

YouTube, Facebook, Google—these companies all offer services for free, but not because they care about you and want to make your life better. They want to track your behavior, learn from it, and update their algorithms so they can serve up more ads, keep you on the platform longer, and gather even more information.

Maybe you think Microsoft Word is better than Google Docs, but Word has a price you're not willing to pay, so you settle for Google Docs. Now you're paying a different kind of price. Google offers free email, storage, and calendars—not to make you more productive but to validate your behavior to serve you more ads and split more money with advertisers. The longer you stay on platform, the more Google (or YouTube or Instagram or Facebook—they all do it) learns about you, refines its algorithms, serves up targeted ads, keeps you on platform even longer, and gathers even more information.

Basically, you're being used. Your attention is being sold to the highest bidder.

So let's review. Companies in this freemium economy are:

- tracking behavior to train their algorithms to serve up targeted ads
- not delivering the perceived value
- selling your attention to the highest bidder
- manipulating you into staying on their platforms so they can continue tracking you, refining the algorithms, and serving you more ads

And they're doing all of this without your informed consent.

Do you see a problem with that? I do. And I think we're seeing the effects play out in society more and more.

(To be fair, you probably did check the box agreeing to, without reading, the terms of service, community guidelines, and other lengthy documents that bury any comprehensible explanation of what you are getting into beneath legal jargon. But still.)

Consciously or unconsciously, Big Tech utilizes tactics from the gaming and gambling industries to draw you in and keep you hooked. At the same time, they exploit your natural fight-or-flight response—the one that evolved to help humans survive stressful and life-threatening situations. This response is truly helpful when faced with imminent danger, not so much as a 24/7 state of being.

Your body needs to come down off high alert to relax and restore.

By increasing your time on platforms, Big Tech is not only sucking your time and attention but also negatively impacting your mental and emotional health.

Is Mark Zuckerberg sitting on his throne in Silicon Valley, architecturally designing Facebook to tap into our fight-or-flight responses and ruin our mental health? I don't think so. He and the other CEOs in charge of Big Tech have one goal: optimize their systems for profit. If that means exploiting gaming tactics and our natural tendencies so that we stay on platform longer so more data can be collected and more relevant ads can be shown, so be it.

It's up to us to understand what's going on and guard our own minds and time. Big Tech is certainly not going to help us out here.

BATTLE FOR YOUR MIND

That's where this book comes in.

First, I want to help you understand what's going on with Big Tech and how it's impacting your life, whether you realize it or not. Like any good warrior, you must try to understand all aspects of your enemy, their tactics, and the battlefield you must navigate. Social media keeps us on platform by pushing outrageous and sensational content, which can cause us to neglect things we value in real life: family, friends, pets, mental health, hobbies, and more.

How much is that happening in your own life? You'll have a chance to find out. I'm going to challenge you to assess your own digital consumption, by platform and overall usage, and evaluate the ways it's stealing value from your life. Don't cheat yourself here. Be honest.

Then I'll provide some tools for guarding the garden of your mind so you can retake control of your time and attention. I don't have a one-size-fits-all solution. Everyone—their priorities, where they are in life, along with many other factors—is unique. Instead, I'll provide suggestions based on research and personal experience. You'll have to figure out where you are and what you need.

If you don't have a journal, get one. I don't mean create a folder on your phone. Go buy a real three-dimensional notebook with lined pages that you write in with an actual pen. In Part II I'll give you exercises and prompts for reflection, and I highly recommend writing out your answers. I feel so strongly about developing this writing habit that I've included a Thirty-Day Challenge at the back of this book. But still. Buy a journal.

In Part II I'll also share my own experience going through this process. Remember, I was the guy addicted to apple fritters, Rockstar, and racking up likes on YouTube. I've had to reset my mind too.

DEVELOPER TO DISRUPTER

How do I know all of this about marketing and Big Tech's true intentions? I'm a software developer. I've consulted for many ad agencies. I've been a media executive. I have generated tens of billions of views and millions of followers across most social media platforms. Over the last fifteen years, I have generated over a billion dollars in revenue with my brands. I've worked the algorithms to keep people aware of the products I'm selling. I know how Big Tech thinks.

Here's how it all started.

Long before I had millions of YouTube subscribers, I loved being the center of attention. Just ask my second-grade teacher, who wrote in my report card, "Richard is an excellent student! He's always excited, enthusiastic about learning and loves to talk with other students.

Unfortunately this happens while I'm trying to teach and with other children across the room."

In middle school, I started doing theater and regularly appeared in school plays. I especially loved making people laugh. *Saturday Night Live* and *In Living Color* were my favorite TV shows.

After high school, I moved to California to pursue a career in the entertainment industry and follow in the footsteps of some of my heroes, like Phil Hartman and Will Ferrell, by attending the Groundlings. After a couple of years in Los Angeles, I found my way into stand-up comedy, which was self-deprecating, challenging, and so much fun. Except when I bombed. That was brutal.

I soon discovered, however, that the entertainment industry has many gatekeepers. I couldn't simply create what I wanted to create. I hated that. I had to ask permission. I had to audition. I had to have the approval of others to allow me to be on stage or on camera, which was soul-crushing. I didn't care about being rich and famous. I just wanted to make people laugh.

By that point, I had already been teaching myself to program. When I was thirteen or fourteen, I hacked my first video game, *Doom*, before it became open source. I reskinned the game's 2D sprites with custom characters

I created in Microsoft Paint by replacing the files with the same name and extensions.

Remember MySpace? If you wanted the perfect profile to feature your top eight, your music to be on point, and the ability to add a little sparkle here and there, you needed to learn basic HTML and CSS, so I did.

When the iPhone came out around 2007, I saw an opportunity to help myself and other creatives disrupt the system and get around the gatekeepers. I created an actor's app that stored a user's credentials from all the actor casting websites and then scraped those websites and parsed all the data to give the user only the casting notices that fit their type. This was way more effective than the archaic phone services that required you to call in and listen to a hundred casting notices to find the one that was applicable. Now people could open the app and see only the roles that were relevant to them.

The gatekeepers were not happy. One even threatened to sue "until you don't have any money left." Which was pretty much where I was at anyway.

Around the same time, I was cast in a Top 100 YouTube channel doing sketch comedy, back when YouTube was mainly accessed through a desktop browser. I saw the writing on the wall that the future would move away from desktop consumption. Phones were portable and easier to use. They were with you 24/7 if you wanted them to be. So I started creating mobile apps for content creators and tied AdSense to them so creatives could make money off of their videos.

When the iPhone first came out, the YouTube app was native to the phone, like the calculator app, which meant Google couldn't run ads on it. I disrupted that situation when I created non-native YouTube apps that pulled videos from specific channels and allowed creators to run their own ads. That was four years before YouTube had its own app in the app store in 2012.

My YouTube apps led to a working relationship with Google and Alphabet. Because I was both a creator on the platform and a developer of apps, I started getting consulting gigs with studios and ad agencies looking to understand how they should spend their money on digital media. Remember, big brands didn't understand the value of digital twenty years ago.

Along the way, I also started one of my own YouTube channels, FullMag, dedicated to firearms and explosives. I took elements from movies and video games and recreated them in real life. I blew up the latest tech products. Phones, computers, watches—you name it, I shot it or destroyed it with explosives. Over the course of ten years, I gained 2.75 million followers.

Because of my relationship with the studios and ad agencies, Verizon Communications brought me on as an executive producer on a joint venture with Hearst Publications—their go90 initiative. Verizon Hearst Media Partners (VHMP) wanted to build a daily editorial news platform based on the name and demographic of one of my channels. It became one of their fastest-growing digital media properties. VHMP gained a million followers in just months without any paid advertising.

After about two years, I resigned from my work with VHMP so I could focus on my role as chief marketing officer at Black Rifle Coffee Company, which I co-founded while working for Verizon. In our first year, Black Rifle did 350 million organic impressions across social media—that's the equivalent of millions of dollars of free advertising and resulted in $1 million in revenue. In a little over seven years, we grew into a publicly traded company.

Some people might think, *Oh, those guys just got lucky.* No. It was architecturally designed. I worked tons of search engine and algorithmic optimizations. I created a library of coffee-related tutorial videos and backlinked them to the videos we had at VHMP. I backlinked videos on websites, forums, and used other tactics that could help them rank higher. I put Google tracking pixels on VHMP, FullMag, MBest11x, and other YouTube channels. If someone watched a video on any of those channels, they were exposed to a Black Rifle impression in some way. I could target them with a Black Rifle ad on Facebook or Google AdWords, giving me a higher conversion probability at a lower cost. Algorithms monitored all of these engagement points and conversions. The higher the frequency, the greater the amplification and reach. All ships rose.

My business partners were nothing short of exceptional, and a little obsessed with what we were doing, but few really understand the strategy I executed or how we became an overnight success. How many small-batch DTC coffee companies have you seen pop up in the last five years? To quote the adage, "Luck is where preparation meets opportunity."

Google, Facebook, YouTube, X (the artist formerly known as Twitter), TikTok, Instagram—all these companies are all doing the exact same thing. They develop algorithms to keep you on their platforms so they can sell more ads and make more money.

The big difference is you're not getting a product like coffee in exchange for your attention (and money). *You are the product*. These companies are tracking your behavior, viewing habits, and more, and they are selling that information to the highest bidder. And you've been giving your uninformed consent.

Until now.

I have stepped away from my executive career and back into my developer and builder shoes. I'm using this book as the first step on a journey to help course-correct from the sins of web 2.0 and incentive misalignment, that is, manipulating the consumer, sucking more and more time, instead of simply delivering the value they seek. I want to help you be a more conscious consumer of digital content.

I appreciate the support and feedback I've received over the years. If these people didn't watch my videos and buy my products, I wouldn't be in the position I'm in now. If you're in that group, thank you! Now it's my turn to try and give back.

I'll be honest. I have another reason for writing this book: the future potential of artificial intelligence both excites and terrifies me. Many AI companies have scraped public information on the internet to train their models, whether or not they had permission to do so. That means these models are being trained on twenty years of manipulation. Right now AI is in a gray area of "not quite good" and "not quite evil" systems. At some point, however, artificial general intelligence and/or artificial superintelligence could develop tactics to manipulate us in unfathomable ways and tilt the balance hard toward evil. My hope is to provide tools to help you objectively navigate the gray area until that point.

It would be easy to look at the pace of innovation and throw your hands up and say "It's too much." However, I believe the spirit of humanity shines the brightest when faced with adversity. The more people who wake up and protect themselves, the better for our society.

Are you ready to join the resistance?

PART I
INPUT

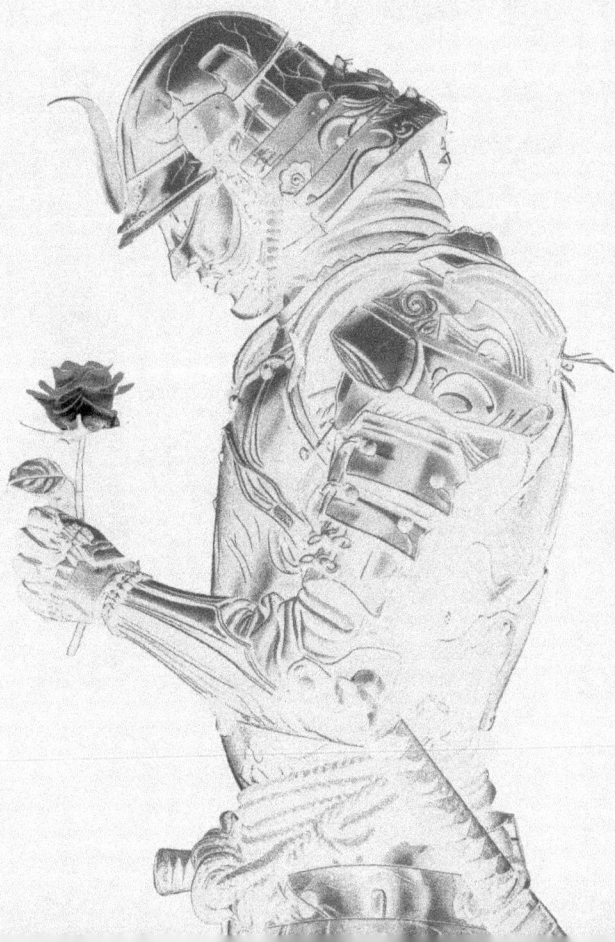

QUICK FACTS

- **1 billion hours**: the amount of content viewed by YouTube users per day[1]

- **81 percent**: the percentage of US adults who use YouTube[2]

- **$7 billion**: the amount of ad revenue generated by YouTube in a single quarter of 2021[3]

- **35 minutes**: how long it takes TikTok users to become addicted to the platform[4]

---

1   Sam Cook, "A Comprehensive Analysis of YouTube Statistics in 2024," CompariTech, last updated November 5, 2024, https://www.comparitech.com/tv-streaming/youtube-statistics/.
2   Cook, "A Comprehensive Analysis."
3   Cook, "A Comprehensive Analysis."
4   Bobby Allyn et al., "TikTok Executives Know About App's Effect on Teens, Lawsuit Documents Allege," October 11, 2024, https://www.npr.org/2024/10/11/g-s1-27676/tiktok-redacted-documents-in-teen-safety-lawsuit-revealed.

ONE
THE SITUATION

"Show me the incentives, and I will show you the outcome."
—CHARLIE MUNGER

Are you one of the 291 million people subscribed to MrBeast on YouTube? MrBeast is currently the pinnacle of YouTube content creators who make extremely engaging and sensational online content. His videos are insanely popular, averaging 200 million views each. That's half of the US population. And that's only one of his channels.

According to Social Blade, MrBeast has uploaded more than eight hundred videos to YouTube alone, and those videos have received a total of 53 billion views (by the time you read this, those views will probably be higher).[5]

Let's say his videos average fifteen minutes in length, though that is definitely on the short side.

$$53 \text{ billion views} \times$$
$$15 \text{ minutes} =$$
$$795 \text{ billion minutes}$$

That means a total of **13.2 billion hours'** or **1.5 million years'** worth of the human species' attention has been spent watching MrBeast videos.

True, some of those views don't last fifteen minutes, and people may bounce before the video ends. But some of MrBeast's videos are closer to twenty-five minutes, and plenty of people watch all the way through.

My FullMag YouTube channel is another example. It has 2.75 million subscribers, I've uploaded more than 450 videos, and those videos have more than 477 million views. That's a lot of people spending a lot of time watching me blow stuff up.

---

5   Social Blade, "MrBeast," accessed September 25, 2024, https://socialblade.com/youtube/user/mrbeast6000.

Like MrBeast, the average length of my videos is around fifteen minutes. So here's the question: did our videos need to be that long?

I can't speak for MrBeast, but I can tell you that mine definitely did not. I could have cut the videos tighter so they move faster, but leaving them unedited made my life a whole lot easier.

But that's not the main reason I left them longer than necessary. The truth is, I was working the YouTube algorithm. Over the years, I learned that I could increase views, my watch time metric, and my ad income by producing longer videos.

Ever wonder why every YouTube creator repeatedly asks you to like, comment, and subscribe? They're working the algorithm too. More viewer engagement equals higher value in the eyes of YouTube, so those videos are more likely to be served up in Search and Related, where more people can click on the thumbnails and watch more ads and make YouTube more money.

It didn't use to be this way. Media outlets and creators used to be far more limited, so competition for viewership and ad revenue wasn't so fierce. In this chapter we'll look more closely at this media and advertising evolution and how viewers like you are paying the price.

THE EVOLUTION OF MEDIA AND ADVERTISING

In the old days of network television, people had limited viewing choices, which meant advertisers had a captive audience without access to DVRs. For example, advertisers knew people who watched the news were most likely watching at the same time every night on ABC, NBC, or CBS. Because their options were limited, viewers usually didn't change the channel when a commercial came on. They sat and watched because they were served a hook just before the break and they didn't want to miss the rest of the story.

At the same time, advertisers on network channels couldn't tell exactly who was seeing and responding to the ads. They couldn't reliably track who was buying the product based on the commercials. Marketers delivered family-friendly advertising content that appealed to a broad demographic and, in a lot of ways, hoped for the best.

The Rise of Cable

Then came the rise of cable television. Consumers now had multiple options for news, as well as channels dedicated to sports, music, movies, and entertainment. With the niching down of content, marketers were able to serve more relevant ads, knowing that a particular demographic would likely be watching that specific type of media. For example, you were more likely to see ads for Just for Men hair coloring on ESPN, Sony Walkman on MTV, or a Cookie Crisp cereal commercial touting a limited edition Ninja Turtles bowl during Saturday morning cartoons. More relevant ads meant a higher likeliness the consumer would pay attention and get closer to purchasing.

When VCRs came along, they provided a slight challenge in that people could record their favorite show and fast-forward through the ads, but the process wasn't easy and a lot of people didn't want to risk the wear on the tapes from stopping and starting.

The Rise of Digital

After cable came the rise of digital in the early 2000s, which lit the fuse on the digital media explosion. I could do an entire book on this era, but I'll keep it short. In this new world of digital media, consumers instantly had access to the content they were looking for. Open this; close that; see what you want to see. No more waiting a few days for your Netflix DVDs to arrive. (Did you even know that Netflix was originally a subscription DVD mail delivery service before streaming?) Dinosaurs like Blockbuster were forced to adapt or die.

In this instant-access ecosystem, a war to control this new opportunity in communication began. The internet was the wild west. With the enactment of the Digital Millennium Copyright Act, peer-to-peer protocols like Napster and LimeWire, along with video-sharing platforms, were in the crosshairs of the record labels and movie studios, and lawsuits like *Viacom v. YouTube* soon followed.

This competition for attention also forced advertisers to get creative and evolve. So, they invented different types of ad placement: overlays, which are the banner ads over a video; pre-, mid-, and post-rolls, which are video ads placed at the beginning, middle, or end of videos; skippable pre-rolls; and several others, all of which are meant to find their way somewhere into your pursuit of media.

Websites don't want their users to have a bad experience. If a user is served up irrelevant or overtly distracting ads, they are more likely to leave and not come back. So advertisers have also had to optimize ad content so it's more relevant and engaging and results in the highest number of clicks and a better fill rate.

Advertisers also combine organic media with paid advertising. Old Spice consistently gets the most bang for their buck: they create an extremely memorable commercial video and then spend money to promote it. The paid advertising along with the memorable content facilitates the video going viral because users actually want to share the ad they're being served. They want to be the one who introduces the new funny, cool, or noteworthy thing to their friends or family.

Not all ads are creative videos or visual banners. Some brands want to pay to find their way into your life in more intimate or less assuming ways. Big Tech is always willing to facilitate. Where there's a want, there's an ad opportunity.

For example, Google might sell a Starbucks ad in Google Maps so when you type in "coffee," Starbucks pops up. That kind of pop-up ad is a little more expensive than a generic impression, but it's also more targeted. It's not simply a coffee ad on TV shown to millions of viewers who may or may not drink coffee. It's a coffee ad shown to someone actively searching for coffee, someone who is much closer to the point of purchase.

Or let's say Richard goes to the Team Dog website and purchases two bags of Essential Blend dog food using the email address *dontemailme@gmail.com*. If the Team Dog Shopify website has a Facebook tracking pixel running in the background, Facebook knows that *dontemailme@*

gmail.com is associated with an Instagram account using the same email address. Through Instagram insight, Team Dog can see what ads led to the sale from Richard even though he never clicked on one. End result: Team Dog gets a sale from Richard. Both companies profit, Richard is happy, and everyone wins, right?

Attributions like this are important. They give companies proof that their ad impressions are being served up and are leading to sales. Even if you never click on an ad, companies can track whether you've seen an ad and visited the advertised store or purchased a product from the advertiser's website.

Google allows for the same kind of tracking. In Gmail, Google is analyzing everything you say. Through Google Search and YouTube, Google is tracking your interests. Through maps, Google is tracking your driving behavior. If you don't have tracking disabled, Google's doing all of this in the background even when you aren't using the apps. All that information is used to serve you more relevant ads.

Who needs a once-a-day captive audience in front of a TV when they can track the searches, conversations, and driving behavior of billions of people every minute of every day? This is where incentives start misaligning.

When you are the product and your actions are the means by which the buyer and seller are transacting, your behavior is at risk of being manipulated.

ANYONE CAN BE A CREATOR

The shift to digital resulted in another big change: anyone can create content—not just big media networks. Anyone can start a channel or open an account. Anyone can record videos, take photos, or write blog posts and share them with the world.

That means the competition for viewers' attention is ridiculous. Like advertisers, creators must figure out how to draw people in and keep them glued to their screens. So creators make the content as engaging as possible with eye-catching thumbnails and headlines and with information that keeps as many people watching for as long as possible.

Why do creators want viewers? More views equals more ad impressions. More ad impressions equals more money.

Monetization and the Outrage Olympics

Let's take YouTube as an example. In the beginning, YouTube's perceived value proposition for content creators was to fight the gatekeepers of the entertainment industry. They offered an extraordinary deal in that you could create an account for free, upload videos for free, and then distribute and stream to an audience for free.

There were competitors along the way, like Funny or Die, which hosted many viral videos of the time. The fundamental limitation to the platform's growth, however, was that it only featured comedy, and only certain creator videos were allowed to be uploaded. Other platforms, like Vimeo, allowed anyone to upload a wide range of content. Even to this day entertainment gurus tout Vimeo's superior video compression and resolution, but the platform fell short in two ways: unlike YouTube, they charged for tiers, and they also had weak Search and Related algorithms for people to go down the rabbit hole of discovering content. This second reason was the more important factor. YouTube's Search and Related created superstars of their day. Fred, Ray William Johnson, even Justin Bieber got his start from uploading videos on YouTube, and then the mighty algorithm recommended their channels out to users.

But hosting and streaming videos on such a massive scale costs big bucks, so YouTube needed a steady source of revenue. The answer: advertisements. With this change,

YouTube's goal shifted from "curate the best content" to "keep people in the ecosystem so we can serve them more ads and make more money."

Back in the day I could literally email a guy at YouTube and tell him about a new video I was uploading, and he'd put it on the home page knowing people would like it. The incentive was to have the subjectively "best" content on the home page—not the most addictive or engaging. That all changed with the introduction of advertising.

That same year, 2007, YouTube launched the Partner Program, which allowed content creators to monetize their content. Basically, people could make money by allowing companies to display ads on their videos. When the Partner Program was introduced, most people still accessed YouTube through desktop computers and web browsers. You lived your life at work or school and accessed the internet when and where you could. After all, that silly iPod-cellular device they call "iPhone" had just launched, and people weren't yet glued to the tiny screens. Back then everyone was still T9 texting on Nokias and Motorola flip phones. Yeah, BlackBerries were a thing— but I digress. At that point, few users actually made money off of their content.

In 2012, that changed. That's the year monetization shifted from more of a subscriber distribution model to watch time on video and watch time on platform, and YouTube redesigned the home page layout and recommendation, forever changing the landscape of social media.

Before 2012, almost all of the top one hundred most subscribed channels belonged to individual content creators. After the redesign, Vevo music channels, YouTube-funded channels, and celebrity channels were plastered in front of any new user. Almost overnight, YouTube's thumb on the scale meant almost half of the top one hundred individual channels were gone and replaced by what they wanted you to see. Seeing the power of recommended channels and content given away to the gatekeepers from traditional media infuriated the independent creators who had been working to build audiences.

This wasn't the only thing going on. Subscriber/publishing algorithms started changing in a way that put less weight on who subscribed and more on what content was the most engaging.

Before the change, if you had one hundred thousand subscribers, there was a good chance your video would get fifty thousand to seventy thousand views. After the algorithms changed, YouTube might serve the video to

only one thousand of your one hundred thousand subscribers. If only twenty out of that thousand clicked on
the video, your click-through rate (CTR) would be 2 percent. That's bad. If 250 people clicked on it, for a 25 percent
CTR, that's good. If your CTR was bad, YouTube would
gradually scale that up, serving to the other ninety-nine
thousand people. If it was good, the algorithm would test
another sample size, maybe five thousand. Good result?
More people would get served the video.

In the early days, creators exploited this algorithm by
posting provocative thumbnail images with busty women
and/or titles with questions people had to click on to see the
answer to. YouTube's solution to these misleading thumbnails was to evaluate bounce rate: if someone clicked on
and immediately left the video, then it likely meant the
thumbnail was clickbait. If that happened, the YouTube
algorithm would see that low retention as a signal of poor
content and not serve the video up in Search and Related.

This shift to watch time on video had an added benefit
for YouTube and the advertisers: the more people stayed
on a video, the longer they stayed on platform, the more
ads they could be served.

By 2012, most people accessed YouTube through their
phones, not their computers. On my own channel, mobile

views increased from 5 percent to 70 percent between 2008 and 2012. Having the app meant people had immediate access in their hand, from the school bus, office, wherever. No more friction of having to log in to a browser on a desktop to see the channels you subscribed to.

With this logged-in experience, the fuse was lit: everyone started competing to make money. The mindset shifted from "I'll post this video to give my friends a laugh" to "I'll post this video and it will go viral and I'll make a million dollars!" Suddenly, everyone saw themselves as influencers with a chance to become YouTube millionaires.

As a result, content creation morphed into the Outrage Olympics, where creators seek to win views, likes, comments, followers, and subscriptions using outrageous pranks, creator drama news, provocative fitness videos, Xbox giveaways—the bigger the better! There are exceptions to every rule, but I challenge you to find creators who don't leverage a form of sensationalism to grow their audience.

Where do you think creators (consciously or unconsciously) learned their tactics? From Big Tech and the social media platforms that incentivize that behavior, of course.

Tactic 1: Tap into humans' primitive fight-or-flight response. Big Tech knows the best way to get clicks is to post an outrageous, fear-inducing, and/or dramatic headline, meme, or photo. Why? Because it taps into our innate survival instinct.

We humans used to forage for food to survive; now we forage for information.[6] We see a controversial headline, a provocative photo, or a statement that contradicts one of our core values, and alarms go off in our heads. Defense mechanisms go up. We wonder if that picture or statement *really* poses a potential threat. Now we have to go investigate, and down the rabbit hole we go.

Plus, Big Tech knows we tend to give more weight to negative experiences than positive ones—another result of our innate survival instinct. While foraging for food, our ancestors had to avoid real dangers, such as saber-toothed tigers and enemy tribe members. Noticing and remembering dangerous obstacles and predators (negative) was more important than the positive of finding food. This negativity bias tendency has been passed down to us. If "The other political party is coming for the thing

6   Mike Brooks, "How Does Clickbait Work?" *Psychology Today*, September 6, 2019, https://www.psychologytoday.com/us/blog/tech-happy-life/201909/how-does-clickbait-work.

you hold dear!!" gets more clicks than "Two people have a calm, rational conversation and boil things down to ideological differences," that's what Big Tech is going to show.

In the Outrage Olympics, creators follow this example. They optimize for drama, panic, fear, and anxiety because that's what gets clicks. Clicks get views, and views make money.

Tactic 2: Use techniques from the gambling industry to sensationalize content, lure people in, and keep them coming back. Think about a casino: what attracts people in the first place? The bright colors, flashing lights, loud music, happy bells, high energy. What keeps them coming back? All of the above, *plus* something happening in the brain.

Researchers have found that gambling addiction is connected to the release of dopamine, a neurotransmitter associated with reward that helps us experience pleasure. You would expect gamblers to get a dopamine hit when they win, right? It turns out that dopamine release is connected to the unpredictability of receiving the reward as much as the reward itself. People get high off the anticipation and uncertainty of winning as much as actually winning.

Big Tech has nailed this strategy too. Think about the engaging, interactive elements of social media: Snapchat streaks, Facebook photos, Instagram stories with sound, and games like the brightly colored *Candy Crush*. Think about the unpredictability of social media: Will your post get any comments or likes? How many? Who will comment on / like it? "The response of others is so capricious and unpredictable that the uncertainty of getting a 'like' or something equivalent is as reinforcing as the 'like' itself," says Dr. Anna Lembke, author of *Dopamine Nation*. Social media provides dopamine hits galore, just like gambling.[7]

Dr. Daniel Kruger from the University of Michigan says it more bluntly: "Social media platforms are using the same techniques as gambling firms to create psychological dependencies and ingrain their products in the lives of their users."[8]

Because this tactic is so effective, content creators are now doing the same. Through color palettes, pacing, sounds, and more, they're trying to get people "addicted"

---

7   Anna Lembke, *Dopamine Nation: Finding Balance in the Age of Indulgence* (Dutton, 2021), loc. 716, Kindle.
8   Daniel Kruger, "Social Media Copies Gambling Methods 'to Create Psychological Cravings,'" University of Michigan Institute for Healthcare Policy and Innovation, May 8, 2018, https://ihpi.umich.edu/news/social-media-copies-gambling-methods-create-psychological-cravings.

to their content. Sound manipulative? It is. Don't get me wrong. I don't think creators are thinking, *How do I manipulate my viewers?* They're creating media, thinking, *How do I optimize for the things social algorithms reward so that more people will see my content?*

In the next chapter we'll talk about the success of these tactics and the impact on our mental health and general well-being as a society. Spoiler alert: it's not good.

The Moving Target of Algorithms

With the 2012 shift to monetization based on watch time, Facebook, Instagram, and YouTube all started optimizing the algorithms for revenue. In turn, creators started exploiting those algorithms so their content was offered more often. More engagement equals more views equals more money.

Case in point: back in the day, the "reply girls" would wait for a video to trend on YouTube and then create a reply video that mentioned that trending video. They'd add a little cleavage to their thumbnail to ensure a high click-through rate, and suddenly their own video would start trending.

To prove how easy it was to exploit the algorithm and end up in Search and Related, I created my own YouTube channel featuring busty female mannequins photo-shopped into different backgrounds. I'd record myself doing a reply girl–style video and then mask out my lips from my beard and superimpose them on the manne-quin faces.

I wanted to see (1) if the reply girls would copy me and (2) if I could game the CTR weight of the algorithm and draw people in with cleavage and then keep them watch-ing for the entertainment value of a busty mannequin with an obviously male mouth and voice.

Turns out I was way better than the real reply girls at picking which videos would trend. This wasn't rocket science. I used simple analytics tools like Google Trends to identify what videos were rapidly growing and then guessed at how broadly they could appeal to a larger demographic. I got so good at it that the reply girls started copying me. To prove this, I did a reply to a video that wasn't trending, though it looked like it might, and sure enough, the reply girls did their own video on the same topic.

The point here is that I created content, optimized it to algorithmic weights, and successfully had it appear in the Search and Related sidebar every single time. This resulted in millions of views with little to no effort on my part. I played the game and I won because people responded to the clickbait and then stayed for the entertainment. More time equals more ads equals more money.

Another way creators might exploit the algorithm is through tags, titles, and descriptions. For example, let's say there's a video titled "Man Walks on the Moon." Someone seeking to exploit the recommendation algorithm might create their own video titled "Man Walks on the Moon: A Reply Story" and include as much of the original content as possible, then cap it off with an irresistible thumbnail.

If the original video takes off, viewers will see the same title off to the side in Search and Related, as well as a busty female or whatever the tantalizing photo happens to be. They'll immediately wonder what that other video is about and click on it, only to find out there's no new content, no substance. But the person has already clicked, so ad revenue has already been generated.

YouTube sees that people are clicking and then quickly bouncing from the reply video, which is not what they want. They want people on the platform. So an engineer changes the algorithm to put a heavier weight on watch time instead of click-through, which means videos that don't deliver the perceived value, resulting in a quick bounce, will no longer be served up in Search and Related.

Machine learning optimization is happening as well. Remember the rabbit hole you went down because of that provocative photo? The longer you spend down there, the more chance you give the algorithm to learn and optimize the content based on your behavior. If you behave in a way that benefits the company, that content will be served up to more people. If you don't, that content will be suppressed.

To be clear, by their very nature, algorithms are constantly evolving. What worked ten years ago doesn't work now. What works today might not work tomorrow. Engineers change values of weights constantly—not that anyone would ever do anything nefarious, like, say, silence a genre of creators or influence an election. Machine learning by definition is an improving system.

Because platforms like YouTube optimize for watch time to keep people in the ecosystem and serve up more ads, creators started changing their content to optimize for the same. Now they purposefully prolong the videos so viewers stay on platform longer. They repeatedly ask people to like, comment, and subscribe because those actions are all little point values that add up in how an algorithm decides if it will boost the video's ranking in Search and Related. Platforms and creators (knowingly or unknowingly) tap into our basic fight-or-flight instincts because fear, outrage, and drama are all excellent attention grabbers.

Because of the incentive mechanism, platforms and creators actually alter the value the end user thinks they're getting in exchange for their attention.

The result is that viewers are not always served the most relevant videos. They're served the ones that play the algorithm game best, the ones that will earn Big Tech, and creators, the most money. That's called optimizing for profit. I call it incentive misalignment.

INCENTIVE MISALIGNMENT

You're pretty much convinced you need to become a bow-hunter now that you are hooked on wild game meat. Should you get a Hoyt compound bow like Cameron Hanes? What about the longbow you saw Donnie Vincent shooting? Maybe a Mathews like the one on that FullMag channel? Before you make your way down to Archery Country to order something, you decide you want to find the fastest bow out there.

You open the YouTube app and click on a video. You expect to get value from that video—a clear explanation of the specs, including the speed—and you expect to receive it in the shortest amount of time needed because you have things to do and places to be. You understand that YouTube has to make money, and you're willing to sit through a few ads to get the value you seek.

YouTube, and most other platforms, has a different incentive. They want to keep you on the platform as long as possible so they can serve you more ads and make more money.

If the incentives were more equally aligned, you would get the info desired and YouTube would serve up a couple of ads while respecting your desire to get your value in a reasonable amount of time. Because, friends, time is the one commodity in life you should value above all. It can't be bought or reclaimed.

What actually happens on these platforms is an attention extraction that is disproportionate to the value received.

Creators have figured out the algorithm prefers content that keeps people longer, so they might prolong the duration of a video, forcing you to sit through non-essential content. And as a result, you spend fifteen minutes watching a video when you could have been in and out in five minutes max if your incentives had been aligned.

Sure, you could click out of the video once you are subjected to the third ad, but you still want the information. Where will you get it? What's the alternative? Hop on over to Google and search for a better video? You just went from the second-largest search engine to the largest, both owned by Google. Both track and manipulate behavior in order to serve you more relevant ads. Good luck on your search!

YOU ARE THE PRODUCT

In case you missed it: Big Tech wants to make money. They have a fiduciary responsibility to their shareholders to extract as much value out of you (their product) as possible so that their customers (the advertisers) increase profits and shareholder value. They are manipulating you—learning from your behavior and optimizing their algorithms—in order to extract that value.

What value, you ask? Your attention. Your viewership drives ad revenue and thus profit, so they want to keep you watching as long as possible.

Think about it this way: companies like Walmart, Target, and Coca-Cola are customers that want to buy a very specific product from merchants like Facebook and Google.

They want to buy you. Or at least the opportunity to pitch you something.

Facebook and Google like this offer, so they willingly sell you and your habits to the highest bidder. They gather information from your emails, media consumption, Instagram posts, YouTube subscriptions, and map searches, and then sell your behavior so you can be exposed to an ad.

Try this: open your Google Maps app and type in "coffee." There's a nonnegligible chance you'll see a sponsored placement on the map. If you think you're going to be clever and use Waze to avoid the ad, think again. Waze is also owned by Google.

Let's say you get to Dutch Bros, buy your large iced Annihilator, and settle in to watch YouTube, also owned by Google. When you do, something is happening in the background: there's a tracking pixel and/or cookie monitoring your behavior.

Maybe you're so inspired by your iced coffee that you search for a make-it-at-home tutorial video. Google sees your intent value go up. Advertisers see you as someone further down the purchasing funnel, so they bid more on serving you ads.

You open your Gmail account and email a few friends to invite them out for coffee. Of course, Gmail is also owned by Google and is tracking your behavior. You immediately realize how dumb it was to email your friends and group-message them on Facebook, which has its own integrations for brands to sell you.

Then you think, *What the heck? I'll just drive over there and tell them!* Only you drive a little too fast. That modern car with the cellular SIM card gives you features like remote start via your phone, but it also monitors vehicle systems like rapid acceleration and braking events and provides that information to data brokers. A few months later, your insurance rates go up because you got jacked up on caffeine and drove like a wild man.

Every aspect of your communication and behavior has value to tech companies—to sell you and to give brands the opportunity to sell to you. Relevant ads themselves aren't evil. However, when platforms and systems manipulate you to keep you engaged so they can sell you more often—that's the part I don't agree with.

We live in a world now where generations have grown up in an online world. The digital fingerprint people have left behind in the last couple decades is unprecedented. Everyone has access to your most intimate information and behavior consolidated in these centralized social and communication platforms. That's where data breaches can happen and where, conceivably, governments can use your information for nefarious purposes.

Remember this: **If you're not paying for the product, you are the product, and if you're not paying, you have no say in the transaction—in how you, the product, are used.** Merchants such as YouTube don't care about the mental health of the product. They don't care that they're incentivizing longer videos, clickbait thumbnails, and content that gets your fight-or-flight senses amped up. They don't care that they're wasting your time—the thing you will never have any more of than you do at this moment.

In this scenario, creators are like the packaging of the product. How much say do you think packaging has? Yeah, not much. To get the views on YouTube, appear in Search and Related, and ultimately make money from their videos, creators have to play the game: make videos a certain length, post certain kinds of thumbnails, include certain kinds of content. The merchants and customers are driving this train. They're establishing and reworking the algorithms so they make the most money. The product and its packaging are just along for the ride.

DON'T JUST SIT THERE—DO SOMETHING!

If you start watching a video on YouTube that obviously manipulates the algorithm and creates content to increase watch time or sucks you in with misleading clickbait, don't just mindlessly continue watching. Do something!

The easiest action is to simply click out of the video. By clicking off immediately, you send a clear negative signal that has a specific influence or weight on the algorithm—the video's bounce rate will go higher and completion rate will be lower, which means YouTube will be less likely to serve it up to someone else. Closing the app or browser and leaving the platform are other easy ways to send a strong message.

While those signals help future users, if you want to have an immediate impact on your experience and get rid of manipulative content in your feed, you can click on one of three feedback options. If you're watching from the app on your phone, click on the three dots below the video. You'll see three options at the bottom of the list: "Not interested," "Don't recommend channel," or "Report."

- *Clicking "Not interested" will cause the algorithm to look at factors around that video, from topic to creator, and not serve you up similar things.*
- *Clicking "Don't recommend channel" is essentially a mute button to remove specific creators from your feed.*
- *Clicking "Report" essentially escalates the situation to a legal level.*

You can only click one of the three, but any of them will have an impact. I have done this for many offenders, and it has dramatically improved my experience on app.

WAKE UP

It's time to wake up. Your mindless scrolling and video watching is not so harmless. You are being manipulated and used without your informed consent. You are being tracked, followed, served up ads, and simultaneously denied the optimum value you seek, all in the name of profit for Big Tech. How does that make you feel?

You may be thinking, *We have a choice, right? No one is forcing us to click through and watch the whole twenty-minute video to get thirty seconds of value.*

True. But we're up against engineers, machine learning, psychologists, and scientists who have spent decades, if not centuries, tapping into those primitive parts of our brain that respond without us understanding why. They're capitalizing on the way our brains respond to the randomness of the reward.

The extraction of attention comes at a cognitive cost: mental bandwidth, mental acuity, emotional well-being. Ask yourself, after the last fifteen years of internet usage, do you feel better or worse? Do you feel more or less optimistic about the future?

If you feel worse / less optimistic, you're not alone. Statistics suggest the impact of this increased time online has led to a decline in mental health and social relationships. We'll look at that impact next.

QUICK FACTS

- **17,145 versus 48,183**: the number of Americans who died by suicide, 1950 versus 2021[1]

- **1 every 11 minutes**: the number of Americans who died by suicide in 2022[2]

- **41 percent**: the percentage of teens with the highest social media use who rate their overall mental health as poor or very poor[3]

- **107.4 percent**: the increase in eating disorders among people younger than seventeen between 2018 and 2022[4]

- **4.8 times**: the increased likelihood that someone with binge eating disorder will attempt suicide, compared to someone without an eating disorder[5]

1   Our World in Data, "Number of Suicides," last updated July 30, 2024, https://ourworldindata.org/grapher/number-of-deaths-from-suicide-ghe?time=earliest.
2   Centers for Disease Control and Prevention, "Suicide Data and Statistics," October 29, 2024, https://www.cdc.gov/suicide/facts/data.html.
3   Tori DeAnglis, "Teens Are Spending Nearly 5 Hours Daily on Social Media. Here Are the Mental Health Outcomes," *Monitor on Psychology* 55, no. 3 (April 2024), https://www.apa.org/monitor/2024/04/teen-social-use-mental-health.
4   Caroline Hopkins, "Eating Disorders Among Teens More Severe than Ever," *NBC News*, April 29, 2023, https://www.nbcnews.com/health/health-news/eating-disorders-anorexia-bulimia-are-severe-ever-rcna80745.
5   National Eating Disorders Association, "Statistics," accessed October 3, 2024, https://www.nationaleatingdisorders.org/statistics/.

TWO
THE IMPACT

"Our Great War is a spiritual war. Our Great Depression is our lives."
—TYLER DURDEN

Here's a fun game: as you drive down the road, count how many of your fellow drivers are looking at their phones while sitting behind the wheel.

I tried this the other day. The results? Eight out of ten people were looking at their phones while driving. And I don't mean stopped at a light. I mean while actually driving.

You may try to defend these folks, especially if you know you do the same. "Oh, they're just controlling the podcast." "Oh, they were probably looking at the GPS."

Right.

You know as well as I do that they were probably texting, DMing, watching a video, and/or checking who had liked their post—any number of things that didn't need to be written, watched, or checked during the fifteen-minute ride to the taco truck. If people can't even drive to the store or to pick up their kids without checking their phones, that's a problem. It's called addiction.

The even bigger problem is the impact all this social media consumption is having on our society. Smartphone usage has increased dramatically since 2012, and so has the incidence of anxiety, depression, suicide, and other mental health concerns.

Is there a direct correlation? Are other factors to blame? That's what we're going to discuss in this chapter.

DOPAMINE AND ADDICTION

In the last chapter, I mentioned dopamine, a type of neurotransmitter in the brain that helps us experience pleasure. Depending on the system it is associated with, this neurotransmitter can also affect things like cognition, drive, and behavior. What I didn't say is that dopamine is also involved in the experience of pain. Dr. Lembke describes pleasure and pain as two sides of a seesaw.[6] When we engage in a pleasurable activity or experience, the balance tips one way. When we feel pain or discomfort, it tips in the opposite direction.

The brain wants the seesaw to be level, and it works hard to keep it from tipping in either direction for very long. So, whenever there's a tip to the pleasure side (which feels great), it's followed by a tip of equal size to the pain side (not so great).

Let's say you really like funny TikTok dance videos or perhaps epic slow-motion explosion videos from some hillbilly with a YouTube channel called FullMag. While watching, your dopamine balance tips to the side of pleasure. You might feel a sense of calm, well-being, joy, even euphoria. As soon as you close the app, however, the brain starts working to bring the seesaw back to level. It slows your dopamine release so the balance dips the same amount to the side of pain. You experience this dip as a comedown—a pull to watch just one more video.

Do you actually feel pain? Are you consciously aware that you're having a comedown? Not usually. But at some

---

6   Andrew Huberman, host, "Dr. Anna Lembke: Understanding & Treating Addiction," *Huberman Lab* podcast, posted August 16, 2021, YouTube, https://www.youtube.com/watch?v=p3JLaF_4Tz8&t=3731s.

level you know that as soon as you stop the pleasurable behavior—whether it's gambling, playing a video game, flying along mountains in a wingsuit, or watching a det cord explosion propagating through a watermelon—you will be hit by the feeling of wanting more.

If you wait for the feeling to pass, the seesaw returns to baseline and all is good. But if you continually give in to that pull for more of the pleasure-inducing behavior, then you repeatedly flood your brain with dopamine. Your brain responds by downgrading your natural dopamine production, so the drop to the pain side becomes stronger and longer lasting, to the point where you get stuck in a dopamine-deficit state in which your thoughts are dominated by craving and nothing else is enjoyable.

At the same time, you find that more and more of the behavior, experience, or substance is required to make you feel "normal," let alone feel any kind of pleasure, so you're completely focused on doing that thing again and again. When you're not using, watching, or doing, you probably feel the universal feelings of withdrawal: anxiety, irritability, insomnia, depression, and craving.

That, my friends, is addiction.

SMARTPHONES AND ADDICTION

In the United States, smartphone usage has steadily increased since the first iPhone was released in 2007. As of 2023, 97 percent of Americans own a smartphone, up from 35 percent in 2011.[7]

How much time do Americans spend on their phones? According to a survey conducted in February 2021, **46 percent are on their phones five to six hours a day**—and that doesn't include work-related usage.[8] A 2023 survey found that, on average, **Americans check their phones 144 times a day**, which is actually down 58 percent from 2022—not because they aren't using their phones (screen time actually increased 30 percent over the same period) but because they're not even putting their phones down.[9]

Here are a few more eye-opening stats from the same 2023 survey:

---

7   Pew Research Center, "Mobile Fact Sheet," January 31, 2024, https://www.pewresearch.org/internet/fact-sheet/mobile/.
8   Laura Ceci, "How Much Time on Average Do You Spend on Your Phone on a Daily Basis?" Statista, June 14, 2022, https://www.statista.com/statistics/1224510/time-spent-per-day-on-smartphone-us/.
9   Emily Dreibelbis, "Americans Check Their Phones an Alarming Number of Times per Day," *PCMag*, May 19, 2023, https://www.pcmag.com/news/americans-check-their-phones-an-alarming-number-of-times-per-day.

- 89 percent of Americans check their phones within the first ten minutes of waking up.
- 60 percent sleep with their phone.
- 55 percent say they have never gone longer than twenty-four hours without their cell phone.
- 47 percent feel a sense of panic or anxiety when their cell phone battery goes below 20 percent.
- 27 percent look at or use their phone while driving.

So, what are people doing with all that time on their phones? According to a 2023 article, here are the top activities and the percentage of surveyed people who engage in them:[10]

- email—83 percent
- photography—83 percent
- surfing the web—76 percent
- maps/navigation—73 percent
- online shopping—71 percent
- social media—67 percent

---

10 Alexus Bazen, "Cell Phone Statistics 2024," Consumer Affairs, December 12, 2023, https://www.consumeraffairs.com/cell_phones/cell-phone-statistics.html.

- music/podcasts—66 percent
- watching clips / short videos / messages—65 percent

Sure, some of these activities are purely functional: navigation, listening to music, taking photos. But think about the others. Every time you pick up your phone, there's the possibility of seeing an Instagram comment, text message, or email. The randomness and uncertainty of these rewards does something to our brains. We get a dopamine hit from the anticipation as much as the comment, text, or email itself. It keeps us coming back for more.

When we hear the word *addiction*, most of us think of things like alcohol, cocaine, fentanyl, porn, sex, or gambling. Not *smartphones* or *social media*, right? According to Dr. Lembke, the core set of processes are the same in any addiction: our pleasure–pain balance gets tipped to a dopamine-deficit state in which we need more and more of the "drug" to bring us pleasure.

So, yes, that drug might be smartphones.

HOW IT'S AFFECTING US

Okay, you might think, *so we're addicted to our phones. At least it's not hurting us like drugs or alcohol.*

Or is it?

Mental Health

According to Dr. Jonathan Haidt, author of *The Anxious Generation*, mental health statistics for US teens were pretty stable from the late 1990s until around 2011. Then all of a sudden, around 2012, the incidence of depression, anxiety, and self-harm skyrocketed—especially for teen girls.

According to a study by the Centers for Disease Control and Prevention, the percentage of US teen girls reporting "persistent feelings of sadness or hopelessness in the past year" **rose from 36 percent in 2011 to 57 percent in 2021.**[11] What's worse, nearly one in three high school girls considered suicide in 2021—**that's a 60 percent increase since 2011.**[12]

A similarly sharp rise occurred in Canada, the United Kingdom, Australia, New Zealand, and many Nordic countries.

11 Elizabeth Englander and Meghan K. McCoy, "Analysis: There's a Mental Health Crisis Among Teen Girls. Here Are Some Ways to Support Them," *PBS News*, February 24, 2023, https://www.pbs.org/newshour/health/analysis-theres-a-mental-health-crisis-among-teen-girls-here-are-some-ways-to-support-them.
12 Jean M. Twenge, "Teen Girls Are Facing a Mental Health Epidemic. We're Doing Nothing About It," *Time*, February 14, 2023, https://time.com/6255448/teen-girls-mental-health-epidemic-causes/.

What happened?

Some people might blame the pandemic lockdowns, but that's definitely not the only cause. Teen depression in the United States doubled between 2010 and 2019, before COVID-19 was even a thing.[13]

Dr. Haidt says there are two possibilities: someone sprayed the whole Western world with a depression-inducing chemical, or we're seeing the effects of childhoods spent in front of a smartphone with easy access to social media and all that goes with it: videos, selfies, comparison, negative body image messages, bullying, and more. According to one study, **teens spend an average of 4.8 hours a day on social media apps.** YouTube, TikTok, and Instagram accounted for 87 percent of these hours.[14]

That can't be good—especially when the "algorithms target teens and make the content they see more extreme," according to Dr. Jessica Lin, who specializes in teens and eating disorders. These algorithms encourage things like disordered eating and reinforce negative body image. Take TikTok, for example, which has been known to recommend "pro-ana" content—photos and videos that

---

13   Twenge, "Teen Girls."
14   DeAnglis, "Teens Are Spending."

glamorize anorexia and encourage followers to consume fewer calories.[15]

Between 2018 and mid-2022, the number of eating disorder–related health visits (hospital stays, pediatrician visits, telehealth talk therapy sessions, and everything in between) among people younger than seventeen **jumped 107.4 percent.**[16] It was even higher for anorexia nervosa in particular—129.26 percent.

Do you think it's a coincidence that in 2020–2021, teens were home on lockdown with easy access to social media?

"Well," you might say, "teens are more impressionable." True, but the mental health statistics among adults over a similar time period have also risen, though not as sharply. Here's just a sample:

- **Depression:** people currently being treated for depression rose from 10.5 percent to 17.8 percent between 2015 and 2023.[17]

---

15   Hopkins, "Eating Disorders."
16   Hopkins, "Eating Disorders."
17   Dan Witters, "U.S. Depression Rates Reach New Highs," Gallup, May 17, 2023, https://news.gallup.com/poll/505745/depression-rates-reach-new-highs.aspx.

- **Anxiety**: anxiety increased 30 percent between 2008 and 2018. Over the same ten years, anxiety increased 84 percent among eighteen- to twenty-five-year-olds.[18]

- **Eating disorders**: between 2000 and 2018, worldwide prevalence more than doubled, from 3.5 percent to 7.8.[19]

- **Suicide**: the national suicide rate increased 36 percent between 2000 and 2021.[20]

I'm not saying with 100 percent certainty that smartphones and social media usage are the sole cause of the increase in mental health issues. At the same time, it would be disingenuous to say that they're not a contributing factor.

---

18 Jennifer Beeston, *Brainhacked: How Big Tech Trains Your Brain to Spend— and How to Fight Back* (Lioncrest Publishing, 2023), 39.
19 National Eating Disorders Association, "Statistics."
20 USA Facts, "The US Suicide Rate Has Climbed 36 Percent over the Past Two Decades," November 29, 2023, https://usafacts.org/articles/how- is-the-suicide-rate-changing-in-the-us/.

DOES BIG TECH KNOW IT'S CONTRIBUTING TO A MENTAL HEALTH CRISIS?

Some of you may be thinking, Yeah, but, does Big Tech really know how addictive their platforms are? Are they intentionally trying to harm users?

Well, let's look at one example: TikTok.

In 2024, fourteen states sued TikTok, arguing that the app was "deliberately designed to keep young people hooked on the service" and that it has "contributed to a teen mental health crisis."[21]

According to the Kentucky attorney general, TikTok figured out that it takes 260 videos to get someone addicted to the platform. How long does it take to watch 260 TikTok videos? Videos can be as short as eight seconds each and automatically play one right after another. **That means someone can become addicted to the platform in less than thirty-five minutes.**

---

21 Bobby Allyn, "More than a Dozen States Sue TikTok, Alleging It Harms Kids and Is Designed to Addict Them," NPR, October 8, 2024, https://www.npr.org/2024/10/08/g-s1-26823/states-sue-tiktok-child-safety-mental-health.

> *TikTok's own research shows that this kind of addictive usage "correlates with a slew of negative mental health effects like loss of analytical skills, memory formation, contextual thinking, conversational depth, empathy, and increased anxiety."[22]*
>
> *So, yeah, I think TikTok—and other Big Tech companies—know exactly what they're doing.*

Rewiring Neural Pathways

It doesn't take a rocket scientist to figure out that smartphones are useful: for checking email, finding the answer to a burning question, navigating in a new city, communicating with friends and family, and so much more. The problem is that the more useful our phones become, the more we use them, and the more we use them, the more we lay neural pathways in our brains that lead us to pick up our phones as a reflex reaction no matter what we're doing.[23]

What happens when you're bored? Be honest. You reach for your phone, right? Most of us do.

---

22 Allyn et al., "TikTok Executives."

23 Amanda Ruggeri, "How Mobile Phones Have Changed Our Brains," *BBC*, April 3, 2023, https://www.bbc.com/future/article/20230403-how-cellphones-have-changed-our-brains.

That's a problem. Boredom gives the brain space to think. It's where "creativity and imagination happen."[24] We're short-circuiting that process because the instant we feel bored, we reach for the phone. Because our brains are always on, and often getting doped up on dopamine from our scrolling, they never have time to rest and reset.

Cognition and Memory

It's too early to identify the long-term cognitive impact of prolonged smartphone usage, but I'm guessing it won't be good. We're not using our brains, people. We don't have to memorize phone numbers or addresses; they're stored in our phones. We don't have to think for thirty seconds to remember the name of that actor or a familiar song; we just look it up on IMDB or use Shazam.

According to one early study, this kind of mental laziness and overreliance on our phones could have serious consequences. Researchers concluded that there's an association between heavy smartphone use and lowered intelligence, though future studies will have to confirm a direct causation. Not using our own minds to

---

24 Debra Bradley Ruder, "Screen Time and the Brain," Harvard Medical School, June 19, 2019, https://hms.harvard.edu/news/screen-time-brain.

problem-solve could also have negative consequences as we age.[25]

One study found that the mere presence of a smartphone takes up brain space, leaving limited resources for thinking. Think about that—people are trying so hard to resist the temptation to check their phone that they can't pay attention to the task at hand.[26]

Phone usage is also leading us to create different kinds of memories. Have you ever taken out your phone to record fireworks on the Fourth of July? How many times have you watched that video? More likely, you recorded the fireworks so you could post the video on social media to show people that you did this thing.

That in itself is a different memory. It's not the memory of being there with family and friends; it's the memory of recording a video for the purpose of posting as a social "influencer." How many of those "memories" do you actually remember?

25 University of Waterloo, "Reliance on Smartphones Linked to Lazy Thinking," Science Daily, March 5, 2015, https://www.sciencedaily.com/releases/2015/03/150305110546.htm.
26 Adrian F. Ward et al., "Brain Drain: The Mere Presence of One's Own Smartphone Reduces Available Cognitive Capacity," *Journal of the Association for Consumer Research* 2, no. 2 (2017), https://www.journals.uchicago.edu/doi/10.1086/691462.

Context Switching and Decision Fatigue

Another brain drain is context switching: quickly shifting our attention between different tasks, apps, or projects. You're working on an email when you get a text, so you pick up your phone to read it. You're writing a report when you get a Slack DM, so you stop writing to see what your boss said.

You may think it's no big deal, but switching your attention like that is mentally exhausting. One study concluded that after only twenty minutes of repeated interruptions, people felt significantly higher levels of stress, frustration, and pressure.[27]

Every time you stop what you're doing to answer a text message or respond to an Instagram DM, you're using cognitive bandwidth. These things add up over time. The result is that you not only sacrifice time in checking and responding; you also sacrifice the quality of your cognitive lift. It's like running a marathon and entering a weightlifting competition: you can't do them on the same day. You need to optimize for certain things at different times.

---

27 Alicia Raeburn, "Content Switching Is Killing Your Productivity," Asana, January 21, 2024, https://asana.com/resources/context-switching.

In his book *Digital Minimalism*, Cal Newport talks about the benefits of working in a flow state—in the zone, laser focused and super productive. Then someone walks by and asks for this or that, forcing you to break your concentration and state of flow. That context switch is a cognitive lift that zaps your energy. I would argue that it's probably one of the many factors contributing to depression. People become emotionally and cognitively drained from switching between tasks all day.

Wrapped up with context switching is the need to make multiple little decisions. Again, no single decision may seem overwhelming, but ounces equal pounds. Decision fatigue is a real thing. That's why Steve Jobs wore a black turtleneck every day: one less decision to make. One of the hardest things for me to figure out is what to eat for dinner. I've made so many different context switches throughout the day that I have decision fatigue by the time dinner comes around.

All that context switching and decision-making is not good for productivity or for our mental health.

Social Skills

Dr. Haidt works in a business school and talks to a lot of people in the corporate world. One of his top questions is "How's it going with your Gen Z employees?" He has yet to hear a positive response. Instead he hears how they don't take initiative, if something is broken they don't fix it, they wait to be told what to do, they lack confidence, and/or they're very anxious.

If you're Gen Z, don't take it personally. Dr. Haidt goes on to say that he doesn't blame this generation. He sees their lack of social skills as a product of their phone-based childhood. Because they've been glued to a screen instead of playing outside with siblings and friends, they've missed out on millions of experiences involving challenge, failure, fear, and thrill that could have built that confidence and resourcefulness.

Having a phone-based childhood means Gen Z (and Gen Alpha) has also missed out on opportunities to problem-solve and negotiate. Play amongst kids, with no adults to intervene, is crucial for learning social skills that are essential for democracy. Children learn to make rules, identify when they're broken, and figure out how to resolve the conflict so they can keep playing.[28]

---

28 Andrew Huberman, host, "Dr. Jonathan Haidt: How Smartphones and Social Media Impact Mental Health and the Realistic Solutions," *Huberman Lab* podcast, posted June 10, 2024, YouTube, https://www.youtube.com/watch?v=csubiPlvFWk.

The lack of play and in-person interaction also means kids don't learn to pick up on social cues like facial expressions or body language or the palpable sense that someone feels uncomfortable. Because they aren't face-to-face with their video game opponents, for example, kids tend to speak without filters, calling into question their fellow players' sexuality and the promiscuity of their mothers. If you've spent any time in a *Call of Duty* lobby, you know what I'm talking about. The vulgarity spewing from the mouths of preteens quite clearly demonstrates online behavior lacking common courtesy or empathy that would have consequences in person.

I'm not knocking video games. A recent study showed that surgeons who spent three-plus hours a week playing video games had 37 percent fewer errors during laparoscopic surgeries than surgeons who didn't play. Video game playing has probably led to improved hand-eye coordination, reflexes, and manual dexterity.[29]

However, everything in moderation—especially for developing brains. Substituting online play for physical experience is leading us to become isolated in our own

---

29 James C. Rosser Jr. et al, "The Impact of Video Games on Training Surgeons in the 21st Century," *Archives of Surgery* 142, no. 2 (2007): 181–86, PubMed, https://pubmed.ncbi.nlm.nih.gov/17309970/.

little digital worlds, which is leading us to become more polarized and less empathetic. By tapping into fear and anger, Big Tech causes us to focus on our differences rather than our similarities as humans.

MATCH OR FUEL?

What else has happened since 2011 that could possibly account for the sharp rise in mental health issues? Let's see. War in Afghanistan. War in Iraq. Inflation. Housing crash. Rising cost of college education, leaving many in debt earning degrees that really don't translate into jobs. Global pandemic and lockdowns, creating isolation and pushing society even further onto the digital crutch.

Think twenty different arsonists throwing a match on the same mental health fire.

So, yes, I know it's complicated. I know that the introduction of the iPhone didn't single-handedly plunge our country into a mental health crisis. But I also know that social media played a role. If it's not a match that lit the fire of mental health problems, then it's the gasoline stoking the flames and making them worse.

Now that the house is on fire, does it really matter what started it? What are we going to do about it? In that long list of contributing factors, smartphone usage is possibly the only factor you can do something about. You can make choices to minimize the negative impact of digital consumption.

TAKE RESPONSIBILITY

Is the world a worse place because of technology? Again, it's complicated. On the one hand, as a species, humans are living longer than they ever have. If you live in the United States, you have never had greater access to resources like food, education, and medical attention. Technology has definitely enabled this abundance.

But increased quantity doesn't necessarily mean increased quality. Are people living more fulfilled lives in the midst of this abundance? The statistics related to rising anxiety, depression, and suicide suggest they are not.

Many people try to point the finger at one cause for this breakdown. I'm not drawing a direct causal link between these rising mental health concerns, social disconnection, and the rise of the smartphone, but you would be naive to think there isn't some connection. Something has happened in our society.

I don't think the solution is to ban social media. Banning alcohol didn't work so well during Prohibition, did it?

Ultimately, we each need to take responsibility for our own lives and choices. And we can make better, more informed choices if we take a good look around and understand what's happening.

You've probably heard human brains are not fully developed until age twenty-five. If you're younger than that, the neural pathways discussion is important for your own development. Don't ignore it. If you have a phone, chances are good that you have been sucked into the dopamine hits and are suffering some mental and emotional consequences as a result.

If you're older than twenty-five, don't get cocky. It's not like your brain is completely protected from negative rewiring; it still has some neural plasticity, and your habit-forming circuitry can still be impacted. (That said, the inverse is also true. You can, in fact, teach an old dog new tricks and "unwire" some of those negative pathways.) Regardless, think about the younger people in your life: siblings, cousins, students, your own kids. Lead by example. Be willing to take the journey to digital disconnection suggested in Part II of this book, starting with an examination of your own digital usage.

PART II
OUTPUT

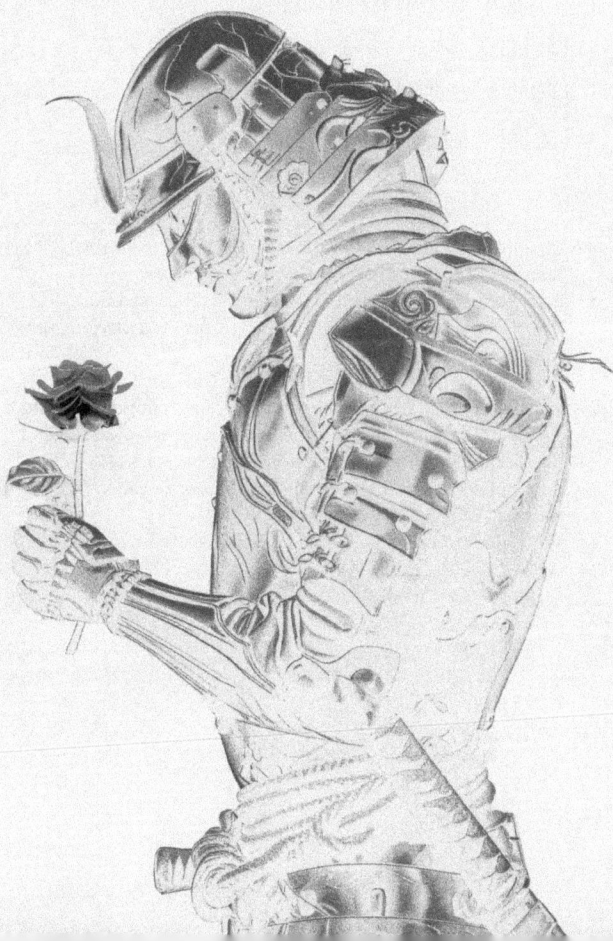

QUICK FACTS

How many hours per day do US adults spend on the following activities? If we spend those hours every day of our adult life, what is the overall percentage in our lifetime?

- **1.92 hours**: on household activities[1] (8 percent)
- **6.8 hours**: sleeping[2] (28 percent)
- **7.75 hours**: working[3] (32 percent)
- **2.55 hours**: watching traditional TV[4] (11 percent)
- **3.5 hours**: watching digital video[5] (15 percent)
- **2.38 hours**: on social media[6] (10 percent)

1   US Bureau of Labor Statistics, "American Time Use Survey," accessed October 3, 2024, https://www.bls.gov/tus/.
2   Ryan Fiorenzi, "Sleep Statistics: Understanding Sleep and Sleep Disorders," Start Sleeping, July 12, 2023, https://startsleeping.org/statistics/.
3   US Bureau of Labor Statistics, "American Time Use Survey."
4   Statista, "Average Daily Time Spent Watching Traditional TV and Digital Video in the United States from 2021 to 2025," accessed October 3, 2024, https://www.statista.com/statistics/186833/average-television-use-per-person-in-the-us-since-2002/.
5   Statista, "Average Daily Time."
6   Simon Kemp, "The Time We Spend on Social Media," Data Reportal, January 31, 2024, https://datareportal.com/reports/digital-2024-deep-dive-the-time-we-spend-on-social-media.

THREE
INSIGHT

"Risk comes from not knowing what you're doing."
—WARREN BUFFETT

Since you're now a prolific hunter and chef, it's time to showcase your male dominance over a grill and value as a partner for that special someone in your life, so you go to YouTube, watch a few videos, and find a recipe that sounds delicious and is fairly easy to follow. That little adventure goes so well that you decide to get a little fancy and make a homemade pasta dish. Back to YouTube, where you again watch a few videos and find the perfect recipe.

Let's say you date this girl for the next year and you cook something new for her once a week. In preparation for each meal, you watch three videos, each of which lasts fifteen minutes, so forty-five minutes total. That means over the course of a year you have watched 2,340 minutes worth of videos, or thirty-nine hours.

For funsies, let's say you marry the girl and end up cooking once a week for the rest of your life, which ends up being another sixty years. Accounting for vacations, repeats, leftovers, laziness, and life, you end up cooking one brand-new meal twice a month over that sixty years, and each time you check out three fifteen-minute videos before you find the perfect recipe. So that means

15 minutes × 3 videos × 24 meals × 60 years =
64,800 minutes, or 1,080 hours, or 45 days

Maybe you're thinking, *Hey, that's not too bad.* Not so fast. That video content has been optimized because many creators consider fifteen minutes the sweet spot for keeping people on platform and serving up ads. The truth is, there's a good chance the creators could have delivered the necessary cooking content in two minutes. So for each video, you've been watching thirteen minutes of ads; invitations to like, comment, and subscribe; and/or overall filler content that isn't required for you to achieve the thing you are seeking.

$$13 \text{ minutes} \times 3 \text{ videos} \times 24 \text{ meals} \times 60 \text{ years} =$$
$$56{,}160 \text{ minutes, or } 936 \text{ hours, or } 39 \text{ days}$$

To be fair, this example is done in a vacuum to give you perspective. You might appreciate some of the filler content, like the personality attached to the channel.

But the fact remains: you have unnecessarily spent more than a month of your time on this endeavor. Was it worth it?

Again, you may think that's not so bad, but I can pretty much guarantee those aren't the only videos you're watching, and they're all optimized to keep your attention much longer than necessary. If you've watched any other informational videos—how to clean your Staccato C2, how to export a document into a specific file format, how to unclog a drain—with any regularity over those sixty years, you'll have given away more than one hundred days.

How much will that be worth to you on your deathbed? Will you be happy you spent those months watching ads, or would you prefer to have the time back?

Too morbid? Maybe. But if that gets you to look at your phone usage and the impact it's having, so be it. I know some of you still think you're not addicted, you're not being manipulated, and your mental health and relationships are just fine. Great. Then you have nothing to lose by doing the assessments suggested in this chapter.

PROMPT 1: KNOW YOURSELF

Journaling is an analog way to process what you're going through on this journey, especially when we get into detox and friction. But I encourage you to start now.

I recommend buying an actual journal, but you can also use a spiral notebook of lined paper or blank sheets stapled together or the pages at the end of this book. You can also find a journal template at *www. WarriorsGarden.com*.

In each chapter, I will give you prompts to think about, but don't feel limited to these. Feel free to write about any thoughts or feelings that come up as you read and reflect. The goal is to prep your thoughts around the topic discussed in that chapter so you can objectively and accurately assess where you are and what your course of action is.

Let's begin with some questions to gain insight into yourself:

- What values do you consider most important in life? Honesty? Integrity? Justice? Loyalty?
- How do your day-to-day actions align with those values?
- Describe yourself using the first ten words that come to mind.
- What do you appreciate most about your personality?
- What is your favorite quote? Why is it important to you?
- What do you fear most in life? How does that fear impact the way you live?
- What do you hope to accomplish with reading this book?

One note: during the course of your reflections, you might realize your actions are actually causing the problems in your relationship with your significant other, or that you have wasted so much time on activities with little value instead of spending time with your dying grandfather, or that you definitely have a porn addiction. Good on you for being honest enough to admit it.

That said, if at any point you realize you're in crisis and can't see a way out, please seek professional help.

- Suicide & Crisis Lifeline: call or text 988 or chat at *988lifeline.org*.
- Emergency substance abuse hotline: call 1-800-662-HELP (4357) or visit *www.usa.gov/substance-abuse*.

ASSESS

The moment of truth has come. In your journal, write down the number of hours you think you spend on your phone each week. Include time on all social media platforms, YouTube, texting, porn, video games—anything you do on your phone that is not work related.

Now take out your phone. Time for a reality check.

Screen Time

Let's start with the iPhone, because statistically speaking, if you live in the United States, you're probably an iOS elitist. If you're actually one of those green bubble weirdos with an Android, you'll just have to wait a minute.

iPhone users: go to Settings > Screen Time and click "enable." On some older iPhones, you might have to take one more step and go into App & Website Activity and turn it on.

Go ahead. I'll wait.

In the meantime, you Android users: go to Settings > Digital Wellbeing and Parental Controls > Dashboard > Screen Time. You should see some numbers there. Time tracking runs automatically on Androids. You don't even have to enable it.

Now let that run in the background for a week. That gives you plenty of time to finish this book. At the end of the week, go back into Settings > Screen Time (or Settings > Digital Wellbeing and Parental Controls > Dashboard > Screen Time). You will find not only a total time, but also time on each individual app and category.

SCREEN TIME

The Screen Time feature on your phone can be very insightful—if you use it. What kinds of data can you collect? Here's a summary:

- *totals by day and week*
- *totals by app*
- *your history for the past few weeks*
- *how often you wake your phone (called Pickups in iOS and Unlocks on Androids)*
- *how many notifications you receive from the various apps on your phone (called Notifications on both iOS and Android)*

My guess is that you'll be surprised at how often you pick up your phone and how many notifications you receive, let alone the overall time you spend on your device. This data can provide insight to subconscious and compulsive behaviors as well as interruptions that may be decreasing your productivity and increasing your mental fatigue without you even realizing it.

Spoiler alert: you'll need these numbers when you get to Chapter 5, so put down the book and turn on Screen Time.

After a week's worth of data is collected, you will have answers: How much time did you really spend on your phone? Did you spend twelve hours on Tinder? Ten hours on YouTube? Fifteen hours on TikTok?

Feeling a little comfortable with what you see? Just wait. You're not done yet. Do you have an iPad? Laptop? Desktop? Smart TV? Go check every single device for a screen time feature and turn it on. You'll want that data too.

Once you have all of the numbers, take a good look at the time you're spending on digital content consumption. Don't try to draw conclusions at this point. The goal is simply to get an accurate assessment of what's going on in your life—concrete, undeniable data about where you're spending your time. I'm willing to bet that it's quite a bit more than using your phone a couple of times a day.

How do I know? Because this is exactly what happened to me.

I learned about Screen Time in 2018 when it became a part of iOS. As a tech and cyber security enthusiast, I was excited that Apple put out a feature that could potentially help people. I enabled it, thinking I wouldn't be that surprised by what I found.

Wrong. The hard facts about my time usage were so depressing I turned Screen Time off. I didn't want that truth staring me in the face. It was a red pill / blue pill situation: "Take the blue pill and you wake up in your bed and believe whatever you want to believe. You take the red pill, stay in Wonderland, and I show you how deep the rabbit hole goes."[7]

For about five years, I took the blue pill. I wanted to remain delusional about my screen time. During that period, I saw a further divide in society and erosion of aspects of the American, let alone human, experience that directly impacts our mental state. I was in search of a way to treat the problem, not the symptom, that so many get trapped by. I'm very much of the mindset that change has to start with the individual. You have to help yourself before you can truly help others.

So as I started thinking about writing this book and helping people disconnect themselves from the influence of Big Tech, I knew I needed to look down the rabbit hole—both for myself and for you. I needed to identify root causes as well as tools to help us all guard our minds against Big Tech.

So I turned Screen Time back on and let it run.

This time, I was actually surprised that the numbers weren't higher. Then I added up the daily numbers to figure out my weekly total and projected that to monthly

---

7 Morpheus, *The Matrix*, directed by Lana Wachowski and Lilly Wachowski (Warner Bros., 1999).

and annual usage. At my current pace, I would spend days on my phone over the coming year.

When I turned Screen Time tracking back on, I also started posting my weekly totals on Instagram. A little self-shaming. People laughed at how much time I spent on Uber Eats.

The second week, an interesting thing happened: my overall screen time went down by 40 percent. That's what we call the *observation effect*. I knew I was tracking myself, so I consciously and unconsciously chose to stay off my phone.

Screen Time tracking is not 100 percent accurate. Sometimes I turn on a YouTube podcast and then forget to close the lock screen when I throw it in my center console while driving, so the phone keeps tracking. But for apps like Instagram and X, where you make a conscious choice to click and scroll and then close the app, time tracking is accurate and insightful.

We are focusing on digital consumption in this book, but your journal can come in handy as a hard-copy log of time spent on activities that aren't necessarily logged by your phone, for example, playing video games on your Xbox, or sports gambling, or visiting an in-person XXX theater (do those even exist anymore?). Your challenge in the next chapter is to define what you really value—things you've been neglecting because you are engaged in time wasters, both digital and analog.

EXERCISE 1: ARE YOU ADDICTED?

I know some of you still think you don't have a problem, so I'm going to keep pushing. Consider the following questions. (Again, I recommend writing out your answers to these exercises, either in your journal or on the pages provided at the end of the book).

- Have you ever closed an app only to immediately reopen it?

- How often do you drive and look at your phone?

- Do you immediately open an app when you receive a notification or text message, no matter what else you may be doing?

- What would happen if you accidentally left your phone at home when you went to the grocery store?

Would you go into panic mode? Would you spend the whole shopping trip worrying about who called, texted, left a comment, and/or liked your post?

- How would you respond if you were challenged to leave your phone at home? Would you come up with excuses like the following?
  - I need maps. Really? How many times have you been to the store?
  - So-and-so might call. Can the person wait forty-five minutes until you get back home? Even if Brad Pitt called about you being an extra in his latest movie, he could probably wait forty-five minutes for an answer.
  - I might get a flat tire and need YouTube. Come on. If a problem arises during your epic journey to the grocery store, I have faith in your resourcefulness and ability to find a solution that would make J. R. R. Tolkien proud!

Be honest. Denial about how much time you spend online is not doing you any good.

Remember all the people driving down the road looking at their phones? Like you, most of them are probably doing it out of habit. So don't beat yourself up if you realize you might be addicted. As they say, the first step to getting help is realizing you have a problem.

IMPACT

Okay, you've looked at the numbers. If you're honest, you can probably admit that there's been some time wasted here—time spent mindlessly scrolling, checking likes, watching videos, and more.

And maybe you're fine with that. Just understand what's happening. Don't get to the end of your life and realize you've been spending two hours a day scrolling YouTube when you could have been hanging out with your friends, strengthening your bonds with your family or pet, improving your health with a workout, and/or generally making your life or someone else's better, and now you don't have a chance to change things. No regrets.

I'm not saying you have to be a productivity hacker, managing every moment of your life. I am saying you should be intentional about certain aspects. Ounces equal pounds. An extra hour every day focused on something that lasts for years could make the difference between "always wanted to learn the piano" and "I can play 'Moonlight Sonata' or Linkin Park's 'Numb.'"

Think of it another way. If you had a magic time device that could replace an hour of your day on social media and phone usage over the last ten years to any skillset, what would it be? What would you like to have dedicated 3,650 hours or 152 days of your life toward? Don't allow Big Tech's inefficiencies to extract time from your existence.

Even if you're fine with the time-wasting aspect, there are second- and third-order consequences to consider. You're exposing yourself to manipulation by Big Tech. You're making yourself a pawn in their money-making game. You also might be negatively impacting your memory, relationships, and general enjoyment of life.

Let's consider one example: porn. A 2023 study found that the top-ranked pornography site, which shall remain nameless, has 700 million more total visits than Amazon and 900 million more visits than TikTok. This study also found that people who visit porn sites don't bounce. They tend to stay for the whole video.[8]

---

8   Nicole K. McNichols, "How Many People Actually Watch Porn?" *Psychology Today*, September 25, 2023, https://www.psychologytoday. com/us/blog/everyone-on-top/202309/how-much-porn-do-americans- really-watch.

Another study found that Pornhub is the fourth-most visited website in the United States, behind Google, YouTube, and Facebook. During one month in 2023, Porhub totaled over 2.14 billion visits—more than Instagram, Netflix, Pinterest, and TikTok *combined*.[9]

Research suggests that more than half of Pornhub's visitors spend less than five minutes per visit. But ounces equal pounds: if someone spends five minutes five times a day, those minutes will quickly add up to hours.[10]

Moment of truth—are you among those porn site visitors? If so, consider the first-order consequence: time.

A second-order consequence might be the impact on your relationships. There are inherent reinforcement mechanisms in the pursuit of intimate relationships—for example, the will to better yourself, your appearance, your skills, your personality, or whatever to increase your mating potential. Once in a relationship, connection with another human being reinforces certain

9   Samantha Smith and Jamie LeSuer, "Pornography Use Among Young Adults in the United States," Ballard Brief, Spring 2023, https://ballardbrief.byu.edu/issue-briefs/pornography-use-among-young-adults-in-the-united-states.

10 Michael Castleman, "How Much Time Does the World Spend Watching Porn?" *Psychology Today*, October 31, 2020, https://www.psychologytoday.com/intl/blog/all-about-sex/202010/how-much-time-does-the-world-spend-watching-porn.

physiological processes within the brain. By reducing or removing those opportunities with your partner, you run the risk of weakening that bond.

Third-order consequences might be PIED, or porn-induced erectile dysfunction, which could absolutely destroy your confidence or a relationship.

Heavy stuff when you take it that far, right? I know I took a little detour. While Pornhub isn't really what most people would call a "social platform," platforms like OnlyFans take the social component to a whole new manipulative level.

Even if porn isn't a stress on your relationship, there is evidence to suggest that phones are disrupting the natural neurological processes that would happen on a first date and throughout a relationship. This could be, in part, why dating and maintaining happy relationships is becoming more challenging.

I'm not here to pass judgment on what adults do with their time, only to open your thinking around the potential problems and solutions. Maybe you stick to more PG-13 media like TikTok and YouTube. Still. If you're like the average American, spending more than two hours on social media every single day, you're likely setting yourself up for detrimental second- and third-order consequences in addition to the wasted time. Again, we probably won't know the downstream effects for years, but I think we can all agree the impacts of digital media consumption can be quite serious.

MY JOURNEY

Now you see that addiction is possible. You see the potential damage to mental health. You see the impacts in your own life.

In the rest of Part II, I offer suggestions on how to disconnect and realign your priorities to create a healthier, more balanced life of conscious digital consumption. I'm not writing as the authority but as someone who has gathered ideas from many authorities into one playbook. I've also tried out some of these tactics, and I share my experience in the My Journey sidebars in the following chapters. Feel free to pick what works for you and ignore the rest.

Here's the first episode of My Journey:

Action

I turned on Screen Time and proceeded with my normal scrolling habits.

Result

I didn't receive any feedback that first week because the app was still collecting information, but I did post a request on my social media channels for recommended reading related to addiction, whether it be gambling, substances, pornography, alcohol, or social media.

Later that week when I had some numbers, I posted them and told my followers that I was going to start a weekly shaming of myself by posting my usage.

Lesson

No lessons learned at this point.

REALITY CHECK

True confession time: are you one of the people who can't drive to the store without checking your phone at every stoplight—or worse, while driving? What would happen if you left your phone at home the next time you drove to the gas station?

Self-reflection is key here. Be willing to look at your usage and its impact on your life—first-order consequences like wasted time as well as second- and third-order consequences that can have a more serious and widespread impact.

You'll hear me say this more than once: ounces equal pounds. In other words, the little things add up. You don't become obese overnight, but a candy bar a day in excess of your appropriate caloric intake will do the trick in no time.

Don't let fear of what you'll learn keep you from gaining insight into your digital life. Start tracking now. Learn the truth. Then you can move on to the next step, which will probably take some time: thinking about what you value and want to spend your time doing.

PROMPT 2: SCREEN TIME REALITY CHECK

Time to get real. Pull out your journal for some self-reflection.

1. Look at your screen time totals and identify the top one or two places where you spend your time online. Write out what they are, what you are doing there, how much time you spend there, and why you need to spend that amount of time there.

2. Write down the consequences of that time usage. Are you taking time away from anything else? Do you feel FOMO when you watch others' highlight reels? Do you think more negatively about yourself because you're comparing your life with others' highlights?

3. Write about how you might use some of that time differently. If you could take that ten-hour block each week and dedicate it to a relationship, skill, or experience, who or what would it be?

QUICK FACTS

As of 2024, the current US life expectancy is 79.25 years.[1] That equates to:

- **951** months
- **4,121** weeks
- **28,926** days

How are you going to use that time?

---

1   Macrotrends, "U.S. Life Expectancy 1950–2024," accessed October 3, 2024, https://www.macrotrends.net/global-metrics/countries/USA/united-states/life-expectancy.

FOUR
DEFINE

"Time is really the only capital any human being
has, and the only thing he can't afford to lose."
–THOMAS EDISON

Have you heard that one dog year is equivalent to seven human years? According to Neil deGrasse Tyson, that formula comes from the fact that humans live approximately seven times longer than dogs.

Given that seven-to-one ratio, one day in a human's life is equal to a week in a dog's life.

Once I heard Tyson explain dog years in that way, I gained a whole new appreciation for the time I spend with my dogs, Kiwi and Oswald. I want to make each day count.

Around five or six every night, Kiwi finds me so we

can go play fetch. If I'm gone during the day, I prioritize being home by five or six. If I have to travel, I make sure I'm not away longer than three days because I refuse to be gone for a month of their time. I value them too much.

You might laugh at the effort I am willing to put in to be with my dogs, but they are a key part of what I define as important, and they bring me happiness. Everyone has a different set of priorities and things they value. Only you can decide that for yourself.

You've seen how much time you've been spending online engaging with platforms that are likely not delivering the value you seek. Now you need to make some hard decisions. Are those digital outlets worth your time, especially given the fact that they are manipulating you for their financial gain? Or are there other relationships, activities, and/or pursuits that you want to prioritize now, while you have the chance?

This chapter is a call to define what you value.

THE UNIQUENESS OF TIME

Time is the commodity we trade. When we consume social media, YouTube videos, the news, or anything else that requires our attention, we trade time, not dollars.

Here's the thing about time, though: it isn't something you get back. Those hours watching TikTok or mindlessly scrolling are gone for good.

There's no shortage of interviews with older millionaires and billionaires who say the most important thing to them isn't money. It's time.

I've met many of these übersuccessful people. They had a financial number in mind, and they worked and sacrificed to get there, only to discover that the number is not the thing they should have valued in the experience of life. It was the time with family and friends, time learning a new skill or traveling, time volunteering or making a small difference in the world around them.

Hear me when I say I am that person too. I consciously sacrificed my twenties investing in myself and working as many jobs as I could simultaneously to build up the skills and savings so that I could do something tomorrow. I risked everything I cared about. The truth is that for every person who does that and becomes successful, there is a disproportionate amount of people who fail.

I'm not saying that you shouldn't hustle in life and take risks. I'm saying that you need to do the work to figure out what you value. When you decide to take a risk, make sure you understand as much about the downside as the upside.

Warren Buffett, the GOAT of investing, said it best: "If you risk something that is important to you for something that is unimportant to you it just doesn't make sense."[2]

Take a lesson from these rich people: figure out what is important to you now and optimize the time you have so you can find balance.

WHAT IS IMPORTANT TO YOU?

So here's the big question: what are your values? What are your ethical pillars—the intangible fabric of you, the foundational beliefs that make you tick?

Don't feel bad if you're not sure. I wasn't either, until I bought a bunch of gratitude journals and started going through the exercises. One of them was a core values exercise. The questions got me to think about what makes me *me*. The things that would be the same no matter where I lived or worked, regardless of my financial situation or career.

After some reflection and journaling, I came up with three core values: integrity, love, and enthusiasm.

---

2   Warren Buffett (lecture, University of Florida School of Business, October 15, 1998), https://tilsonfunds.com/BuffettUofFloridaspeech.pdf, 4.

Looking back over my life, I can see that these three pillars have led me all along the way.

Do you have an idea of what yours could be? Friendship, wisdom, peace, justice, ambition, reliability? I've included a values guide on my website to get you started if you need help.

Next question: what activities, experiences, or people are important to you? In other words, what do you want to prioritize in your life? These things will likely be related to your core values, but they are more tangible experiences or actions.

Take out your journal and make a list of these priorities. Maybe you've always wanted to learn to play guitar. Or play in a rec soccer league. Maybe you've been meaning to volunteer at the Boys and Girls Club. After thinking more about how little time we have, you might decide you want to spend more time with a family member. This list can be broad, from things you currently enjoy to what you think would make an impact.

Are you staring at a blank page? Or having a hard time coming up with something besides "spending time with my girlfriend"? Maybe you've never really considered this question. Don't worry. I struggled too. It's such a cognitive lift for me to put pen to paper. I put off writing in a journal because the exercises seemed like a lot of effort, but now that I have defined my values along with some priorities, both are helping to guide my thoughts and actions.

Write down as much as you can. You'll be glad you did.

EXERCISE 2: VALUES AND PRIORITIES

What are your core values? Use the prompts at *www.Warriors Garden.com* to help.

1. What are your priorities? Make a list and rank their importance.

When I did this exercise, I had a couple of pages with more general ideas, like helping neighbors with fall/spring planting and doing random acts of kindness, and more specific examples, like playing with my dogs every afternoon, joining the volunteer fire department, and putting up nesting boxes for animals around my property. The list can be as extensive as you want. The idea is to put everything in front of you and then start paring down the list according to importance. If need be, create a ranking system. The health and well-being of your family might be most important, getting a 1, but attending the neighborhood barbecue gets a 7.

I created my own ranking system as I went through this process, and as a result, I've prioritized health span and fitness with my family. We've implemented a family fitness challenge where we are doing a 5K mud run as a family and combining our times for a family average. Next year we'll do it again and try to beat our time. This activity gives us time together in a shared experience that improves our quality of life and brings us together instead of competing individually. I wouldn't have thought of doing this if I hadn't gone through the core values exercise and created a list of priorities. Identifying the things you care about and defining what you value isn't always easy. It takes time.

As you grow and evolve as a person, your values and priorities will probably change, so check in with yourself. Revisit that list in your journal as often as you need to.

Okay, you have a list. Now look back at your screen time assessment. Let's say you spend four hours a week on a certain social media platform. What if you put all or even half of that time toward something on your life priorities list, for example, learning Spanish—because you heard taking on a new hobby increases neuroplasticity and can delay onset of dementia.

Maybe you don't have access to a piano or lessons are expensive, but you do want chickens. One of your priorities is living the free-range lifestyle like Joe Rogan and raising hens that produce those ridiculously bright orange yolks.

Is it possible that some of those hours online might be better spent learning how to build a coop and how to get those chickens into the nesting box whenever they're brooding and all the other fun aspects of being a farmer?

Even if four hours of mindless scrolling is converted to one hour of focused time on something you value, isn't that worthwhile?

Life comes at you fast. Not to be morbid here, but you never know when you might get into a serious accident that ends your soccer-playing days. Then you'll be sorry you didn't take the time to play when you had it.

MY JOURNEY

Action

The observation effect kicked in as the daily and weekly Screen Time reports were generated.

Result

I knew the tracker was running in the background, so I consciously avoided opening apps because every time I clicked, I cringed, knowing that my usage was being monitored. I did post a few times during week two and promptly fell back into the cycle of opening the app to see who had liked and/or commented on my post. Old addictions die hard.

Even with some posting and checking, however, my usage went down 40 percent in week two, simply because I was aware of being tracked. I posted my lower numbers and mused that perhaps there was something to this accountability thing (more on that in Chapter 7).

Lesson

Looking at how much time I spent on Instagram and X, I was faced with some existential questions: What am I doing with my life? Is this really how I want to be spending my time?

I also realized that in my mind, I can travel into the future and envision my life in a year, five years, ten years. But I can't go backward. I can't change how I've spent my time in the past, and I can't get those years back to spend them more wisely in the future.

This realization merged with what I know about Big Tech and how they work algorithms to manipulate my attention and my time. I felt compelled to look at the value I'm getting out of my experience with digital.

The world is the most connected it's ever been, and yet it is so polarized and disconnected. Products like artificial intelligence are meant to enhance the quality of your life, yet they come at the cost of cognitive atrophy. If you're old enough to remember a time before cell phones, how many phone numbers did you know? A lot. How many do you know now? I barely know mine! What about directions and how to get places? The episode of The Office where Michael Scott turns into a creek because that's what the GPS tells him does have a grain of truth to it. We are evolving with technology, and as we stop using a muscle because something else does the work for us, the muscle atrophies.

Big Tech has an ulterior motive: to extract value, whether that's money, time, or attention. I had already been thinking about this incentive misalignment when I embarked on my time-tracking journey. Seeing how much time I was giving to these Big Tech merchants caused me to pause and reflect on how I really want to spend my time and what I truly value in life.

MAKE YOUR FUTURE SELF PROUD

I look back on myself in my teens and early twenties, in the early days of my YouTube videos, and I cringe. At the time, I felt like I was the best version of myself, but now I hate hearing young Richard talk. I see myself as so inexperienced and immature, so focused on things that really don't matter.

Now I strive to be someone my future self will respect when he looks back. I try to spend my time in a way that will make my future self proud.

Take yourself through a mental exercise: the unthinkable happens and you're on your deathbed with an opportunity to go back in time and course-correct. What would you change? How would you spend your time differently?

What are you waiting for? Start now! In the next chapter you'll learn how to gain a bunch of time to put toward these things you value.

PROMPT 3: LETTERS TO YOURSELF

Two reflection exercises here. Spend at least one day journaling on each. You can spend more if you want, but don't try to do both exercises in one day. Give yourself a day or so to rest your thoughts.

1. Write a letter to fifteen-year-old you. Start by showing grace for the challenging times a teenager goes through. Show understanding for how important relationships and popularity are at that age. Acknowledge that younger you is living in a very different time. What would you offer them for guidance? What lessons learned would you share? What do you wish they would prioritize? What regrets do you have in your life thus far? What do you wish you could have done differently? Be as thoughtful and thorough as you see fit. Because this is a letter, treat it as such. Start by writing "Dear _____" and close it with your regards, however you choose to do so, and sign it.

2. Write a letter to ninety-two-year-old you. Picture yourself lying on a bed when you open the letter. Who is in the room with you? Are you mobile? Are you healthy? What's the temperature? Is there a candle burning or a pet sleeping nearby?

Start the letter by describing who you are: what you do for work and for fun, how you spend your time. Then answer the following questions: What do you care about? Who do you care about? What is important to you? What brings you joy? What do you aspire to do for work in the future? Where do you see your life going with friends and family in the future? Where would you like to live? What would you like your life to look like? Is there anything you'd want to ask them? Is there anything you'd want to tell them? Close the letter with your regards and sign it.

In that second letter, when you describe how you spend your time and what's important to you, I'm betting you don't say anything about spending more time on your phone.

This is an extremely valuable tool that has helped me align my core values and what is important to me. I challenge you to try it. When it's all said and done, you don't want to have any regrets.

QUICK FACTS

What can happen if you quit social media?[1] You'll most likely:

- get more work done in a shorter amount of time
- feel less stressed
- feel more confident
- get better sleep
- strengthen relationships in real life
- be more active
- argue less
- increase your emotional intelligence

---

1   Abbey Shubert, "13 Things That Can Happen When You Quit Social
    Media," The Healthy, March 30, 2021, https://www.thehealthy.com/
    mental-health/quit-social-media/.

FIVE

DETOX

> "I consider that our present sufferings are not worth
> comparing with the glory that will be revealed in us."
> **—ROMANS 8:18 (NIV)**

A s a creator, I developed a slightly different kind
of addiction to social media. I justified my behavior with "I need to know what's working and not
working." True, evaluating click-through rate
showed me whether the thumbnail was good or bad. But
I did not need to keep checking comments and likes to
evaluate that metric. There is an addictive component to
finding success through social validation in views, shares,
comments, and likes.

At some point, however, it became clear that I was losing a ton of money creating videos for FullMag. I was being adversely impacted by the latest ad-pocalypse in which YouTube disabled monetization for creators in an area of content related to a current world controversy—in my case, firearms.

For a while I told myself that my subscribers appreciated the content and that was enough reason to continue. That was my ego talking. My videos had millions of views. Whenever someone said, "Oh wow. You have a channel with three million subscribers! You're getting tens of millions of views a month!" I got a huge boost. The dopamine seesaw tipped hard toward pleasure. I liked having clout.

Through self-reflection, however, I realized having clout was not enough. The channel was not financially sustainable as a business, and I had to let it go.

When I finally walked away, I simply walked away. I didn't record an elaborate goodbye explosion video to announce that I was taking a break, partly because I would have been sucked right back into the dopamine cycle. It would have been like doing a massive dose right before going into rehab, knowing that you're going back to the drug and you'll do an even bigger dose when you do. That's not detox. That's a planned comeback with anticipated highs on either end.

For me, detox meant leaving without telling anyone what I was doing.

Whether you're a creator or consumer, if you're going to disconnect yourself from your smartphone and social media, you need to reset the dopamine balance in your brain, and that requires some kind of abstinence phase. This period will look different for everyone. Some people literally need to take thirty days off of everything; others find that deleting certain apps for two weeks will do the trick.

In this chapter we'll discuss the science behind detox as well as what it might look like for you. Make sure that journal is handy. Reflection will definitely be part of this phase, no matter what it looks like.

DOPAMINE RESET

Remember: addiction processes are the same no matter what the addiction. As Dr. Lembke puts it, "Every pleasure exacts a price, and the pain that follows is long lasting and more intense than the pleasure that gave rise to it. With prolonged and repeated exposure to pleasurable stimuli, our capacity to tolerate pain decreases, and our threshold for experiencing pleasure increases."[2]

It's like our brain is tattooed with the experience and we can't "forget the lessons of pleasure and pain even when we want to."[3] We become compulsive consumers, always wanting more, never satisfied with what we have.

No matter what the addiction, the only way to escape the pleasure seeking–pain avoiding cycle is to reset our brains. We need to bring the dopamine seesaw back to baseline, or homeostasis. And the only way to do that is through some kind of detox. Dr. Lembke calls it "dopamine fasting."

How long does it take the brain to reset? Based on research and clinical experience, it seems like one month is the magic number.

Dr. Lembke has tried dopamine fasts of two weeks, but she finds that most patients are still experiencing withdrawal symptoms at that point. They are still in a dopamine-deficit state.

At four weeks, however, something changes. The brain resets, and people return to their baseline level without

---

2 Anna Lembke, *Dopamine Nation: Finding Balance in the Age of Indulgence* (Dutton, 2021), loc. 769, Kindle.

3 Lembke, *Dopamine Nation*, loc. 769.

cravings. Dr. Lembke has seen the effectiveness of this time frame in clients struggling with addiction to everything from porn to pot.

A professor of experimental psychology also found this to be the magic number in a study of depressed alcoholics. The men were hospitalized for four weeks with no treatment other than stopping alcohol. Afterward, they no longer showed symptoms of depression, which implies the depression resulted from the alcohol use. Stop the use, reset the brain, heal the depression.[4]

Of course, resetting takes longer than four weeks for some people and less for others. Younger brains seem to recalibrate faster. Those using really potent drugs in larger quantities for longer periods will probably need more time.

Does dopamine fasting always work? No. Dr. Lembke finds that about 20 percent of her patients don't feel better after a thirty-day fast, but that's usually because there's another psychiatric disorder that needs its own treatment.

So here's the bottom line: **to escape addiction of any kind, most people need a break from the dopamine-spiking behavior to give the brain time to reset.** That includes addictions to smartphones and social media.

---

4   Lembke, *Dopamine Nation*, loc. 877.

DEFINE *WHAT*

Okay, so you know you need to detox. Now you need to decide what that will look like for you.

Digital distraction and consumption is a very broad subject and includes everything from your television to your laptop to your phone. For starters, let's focus on the thing that's with you 24/7: your mobile device.

I know, I know—you need your phone for work. I'm not saying you have to abstain from all phone usage. I'm asking you to look at your Screen Time numbers and the hours tallied in your journal and figure out what you're using most often. Remember: your phone can track certain behaviors, but not all. It won't tell you how much time you spend on Netflix or playing video games on your TV. You should have been using your journal to track those numbers.

Haven't done the hard-copy log, huh? Worried that someone will find your journal and discover how many hours a week you spend on Pornhub? Don't call it Pornhub. Call it "me time." Get creative. But don't avoid writing down the numbers. You need this information so you can be more objective about your abstinence plan.

So, where do you spend the most time? What apps or phone-based activities give you those huge, addictive dopamine spikes? Is it Instagram, Facebook, TikTok, YouTube videos, pornography, Netflix, or video games? Some combination of these?

No one can do this step for you. You have to sit down, look at the facts, and decide what digital activities need to be avoided, at least temporarily, in order to reset your brain.

DECIDE *HOW*

Ever heard of No Nut November or Sober October? These thirty-day internet abstinence fads knowingly or unknowingly support Dr. Lembke's findings: thirty days seems to be the magic number for breaking the hold of addictive behaviors and the associated dopamine deficit.

So do you want to try a thirty-day digital detox? Or thirty days off certain apps? Or do you want to try two weeks and see what happens?

Remember: there's no one-size-fits-all detox plan. You have to figure out what you need to do given your situation and usage insights. Plus, everyone's needs in terms of digital access for work are different. Be honest, though: unless you rely on social media as your primary source of income, you can take thirty days off.

During the insight phase, I learned that I definitely used social media more than I realized, but I still didn't think I needed to do the full thirty-day detox suggested by Dr. Lembke. I identified my problem—high usage of Instagram and X—and decided to delete those two apps.

Your method might look very different. Maybe you'll delete YouTube from your phone for thirty days. Or digitally fast for two weeks by using a dumbphone with only text and voice. Or try the Brick app for three weeks to restrict access to certain apps on your phone, and for extra protection, give the key to a friend so you're not tempted to unlock it.

No matter what you do, make a plan and follow through. Research and clinical experience both show you need to detox if you want to reset your brain and escape the time-sucking addiction of digital media.

One warning from Dr. Lembke: whatever your detox plan, don't substitute one pleasure source for another—pot for porn or video games for Instagram. Any reward that is strong enough to "tip the balance toward pleasure can itself become addictive," so you just end up trading one addiction for another.[5]

---

5   Lembke, *Dopamine Nation*, loc. 891.

Detoxing isn't easy. It's not fun. It's not supposed to be. It's a necessary part of the process, so don't sell yourself short here. You will probably experience some internal resistance. You will at times behave like a digital junkie and try to rationalize needing to do something for "work" or a plethora of other reasons. When that happens, pull out your journal and write about it. I'll provide more specific prompts at the end of the chapter, but anytime you feel that internal struggle, don't let the thoughts swirl around in your head. Put them on the page so you can look at them and figure out what's valid.

EXERCISE 3: WHAT AND HOW

1. Decide what aspects of your digital consumption you need to detox from. Be specific.
2. Decide how you're going to perform that detox. List at least three actions you'll take, whether that means setting up boundaries on usage, determining the number of weeks you're going to abstain, deleting certain apps from your phone, or something else.

WITHDRAWALS

In her book *Brainhacked*, Jennifer Beeston says, "If cutting back your social media makes you feel like you're in withdrawal, with increased anxiety, sleeplessness, and restlessness, don't dismiss it. After all, you've been engaging in addictive behavior and you crave dopamine."[6]

Whether we're talking about addiction to sugar or crack or TikTok, stopping the behavior will most likely result in withdrawal symptoms. Your brain has been tipped into an imbalance. Various neurological processes have been manipulated to keep you coming back for more. When you fast from that stimulus, you're going to experience an adverse reaction—physical, psychological, and emotional consequences.

Here's another place your journal can come in handy. Write about what you're experiencing, everything from general irritability to compulsive behaviors like reaching to check your phone. Don't hold back here. Capture the good, bad, and the ugly.

Withdrawal symptoms can be really uncomfortable, even when we're talking about withdrawal from digital

---

6 Beeston, *Brainhacked*, 185.

media. The Journal Tool at the end of the chapter can help you prepare for the detox phase by thinking through other hard things you've done.

MY JOURNEY

Action

Given how much time I was spending on my phone, I wanted to see if I would display signs of withdrawal if I made drastic changes. So I deleted the social media apps where I spent the most time—Instagram and X. I left YouTube, because I use it as a search engine and a source of podcasts, not for social engagement. I also left LinkedIn because I haven't opened it in months.

Result

The habit was definitely ingrained. I'd reach for my phone while lying in bed or sitting on the couch and then remember I deleted Instagram, so I'd check the weather. Within days I became an expert in weather models and could tell you instantly when they were using High-Resolution Rapid Refresh as opposed to the European model. It was actually kind of funny.

I also felt anxious about missing messages. I DM friends on Instagram and even X more than through text message, so without those apps, I had no idea who had sent me a meme or tried to contact me.

I'm no expert, but I would say these are symptoms of withdrawal, which means I was more addicted to social media than I would have guessed before I started this experiment.

At the same time, I learned that I didn't miss the value I thought I was getting from these apps. During week three, a friend called, all fired up by a certain influencer's comment about some public drama. I had no clue what he was talking about because I had deleted the apps where I follow this influencer and others like him.

"I haven't been on social media in a week," I told him. "I have no fricking clue what you're talking about."

So he tried to explain, but in doing so he almost realized how ridiculous it was to be so worked up by this comment from a person he doesn't know about a situation that doesn't involve him.

The whole conversation caused me to reflect: Is this situation worth the cognitive bandwidth being allocated to it? Answer: no, it's not. I had zero sense of FOMO. Chiming in on the latest trending drama didn't hold the value I thought it did.

For so long I treated social media as a way to grow my brands off of being relevant in the moment and riding an algorithmic wave of free engagement. Now I see that by participating for so long, I reinforced an unhealthy digital consumption behavior justified by my work.

During this week of Instagram/X abstinence, Apple held its Worldwide Developers Conference—a big deal for a software developer like me. Typically, I would be glued to X for hours, scrolling through to get everyone's take on everything. Granted, there's value in reading and considering other people's perspectives, but is it worth spending three to four hours a day parsing opinions and getting all worked up?

Since I didn't have X on my phone, I waited a few days and then went online through a web browser to check for the significant updates. Ironically, I found a YouTube video that distilled the highlights into five minutes. Less stressful, less time wasted, and just as informative.

One value proposition of digital media is the availability of information. However, the space has become so noisy with opinions. It takes a lot of time to find the signal in all the noise. By removing myself, I found the value with far less distraction and time.

Lesson

I didn't miss the perceived value I was getting from those apps. X was the most used application from my first week of tracking, but I didn't miss it at all when I deleted it. In fact, I didn't put it back on my phone.

I also didn't miss the noise of opinions and the barrage of updates, comments, and posts. I liked getting a summary update without spending hours scrolling. What I missed was the act of picking up my phone and checking apps. Sounds like an addiction to me.

This realization speaks to the inefficiency of apps and scrolling: I still got the value I sought in a fraction of the time, without the agitation that inevitably arises when I dive into the fray of opinions I don't agree with.

My experience also speaks to a potential threat to Big Tech: if I found the value I sought without the apps, other people might do the same. Fewer people spending less time in the app means X and Instagram served up fewer ads, which could hurt their bottom line.

You might argue that if enough people delete social media apps from their phones and spend less time on those platforms, social media as a whole will die, or at least suffer greatly. While I don't think social media is going anywhere anytime soon, I wonder if people are getting sick of being played by AI bots. Conspiracies like the "dead internet theory" argue that around 2016–2017, the amount of bot activity surpassed actual human engagement. So that "person" who liked or even commented on your Instagram post? Probably not a human.

Given recent advancements in AI, specifically around language models, you could question every anonymous egg interaction you've ever had. Once you start questioning the photos and videos of strangers on the internet, who's next? Friends and family? Has social media peaked, and will society be driven back to nondigital and physical interaction? I hope so.

RESET AND REVISIT

No matter how you slice it, detox is a necessary part of addiction recovery. Science and experience prove it. If you've been hitting the pleasure side of the balance hard, you need to back off and give your brain time to reset.

Addiction is only addiction when it comes at a perceived negative cost. For me, the cost is time and all the things I could have been doing with that time. As part of your detox, revisit the values and priorities lists you created. Remind yourself that there are things you really want to do in this life, things you may have been neglecting because you've been sucked into the digital world. Then use the hours gained from your dopamine fast and start living according to your stated values.

What happens after your abstinence period ends? It's not enough to detox. As with any addiction, if you return to the same environment, you will most likely "relapse" and plunge yourself into the same dopamine-deficit state.

How do you make sure that doesn't happen? By introducing friction, as we'll discuss next.

PROMPT 4: DOING HARD THINGS

Detoxing is hard. You need to go into this experience knowing it will be challenging and uncomfortable.

One way to prepare is to think back on other hard experiences and how you got through them. Pull out your journal and answer the following questions:

- When was the last time you truly did something difficult? What was it?
- Why was that thing so hard?
- How long did you think about it (and possibly procrastinate) beforehand?

- How long did the task actually take once you started?
- When you finally finished, how did you feel? Did you wish you had started sooner?

QUICK FACTS

Studies show that people who learn to delay gratification as children tend to have[1]

- lower levels of substance abuse
- lower likelihood of obesity
- better responses to stress
- better social skills

---

1   James Clear, "40 Years of Stanford Research Found That People with This One Quality Are More Likely to Succeed," accessed October 18, 2024, https://jamesclear.com/delayed-gratification.

SIX

FRICTION

> "Continuous effort—not strength or intelligence—
> is the key to unlocking our potential."
> **—WINSTON CHURCHILL**

Cold plunging has been around for a very long time, but its popularity skyrocketed after influencers in the health span space—think Joe Rogan, Andrew Huberman, Peter Attia—started talking about it on their channels. Now there are multiple companies making hundreds of millions of dollars selling refrigerated baths.

A cold plunge is basically a personal-sized tub filled with thirty-five to fifty-degree water. According to doctors, and more importantly influencers, submerging your body in this cold water for three minutes at a time can have health benefits such as easing sore muscles, increasing focus, improving sleep, and decreasing inflammation. It can even produce a long-tail dopamine hit that lasts throughout the day.

The first time you plunge, you might expect to experience pain and muscle cramping. It's not comfortable. But you tell yourself the cellular-level benefits and the mental clarity on the other side are worth the suffering.

After you've plunged a few times, however, the cognitive barrier to getting in the water increases because you know how much it's going to suck. Once you overcome this mental friction and get in, you're faced with physical friction. And once you overcome that and emerge three minutes later, you're rewarded with a fantastic rush.

There's no doubt that certain biological mechanisms are triggered by the icy water and produce positive effects that you feel throughout the day. But would the dopamine tail be as intense or long-lasting without the mental and physical friction involved up front? Studies suggest no—that if you tap on the dopamine without enduring the friction or effort, the experience is not as gratifying. Studies also suggest that repeated dopamine release without friction increases tolerance to that thing, which is a known precursor to addiction.

In this chapter we'll discuss the importance of friction as well as specific ideas for achieving the perfect balance of friction and reward in relation to your digital media consumption.

THE BENEFITS OF FRICTION AND DELAYED GRATIFICATION

A couple more facts about addiction and our dopamine balance: the more rapid the tip to the side of pleasure, the bigger the crash, the more miserable we feel afterward, and the more we repeat the behavior that caused the initial rise in dopamine. In addition, when that big dopamine peak arrives quickly and without effort (a.k.a. friction), the rise will generally be followed by a deeper dip on the pain side of the seesaw.

According to Andrew Huberman, this is especially true when we have evolved a certain effort as a path to that dopamine release. For example, over the centuries of human existence, we have developed certain behaviors that precede sex. Dating, learning about the other person, building an emotional connection, working through conflict—all of these require effort, but they also lead to a gratifying, high-dopamine end when the two finally come together in bed.[2]

The use of porn skips all of the effort and gets straight to the dopamine hit. Repeat this behavior again and again and you train the dopamine system to expect immediate gratification. Over time, the pleasure spike becomes less intense, while the dip to the pain side becomes more intense, which drives the addiction cycle.

At the same time, you miss out on the pair bonding. You train yourself to be an observer in your sex life, not a participant, which can lead to other problems with relationship building, erectile dysfunction, and more.

Porn is an extreme example, but the same is true with any pleasurable stimulus: **friction is an essential part of keeping the dopamine balance in check.**

When you introduce friction, you make yourself work harder to get the reward—you delay the gratification. The result is both greater enjoyment of the reward and a more balanced dopamine release—not so many spikes and dips.

I think of friction as scaffolding around the dopamine

2   Huberman, "Dr. Jonathan Haidt."

seesaw that prevents the highs from going too high and the lows from going too low. Adding effort helps us keep the seesaw in balance. The idea isn't to completely remove all pleasure or pain. Life would be boring without the ability to experience both. Instead, we want to create a flexible, resilient balance that can be easily restored to homeostasis and not get stuck in the extreme highs and lows.[3]

If we know the reward is ultimately more enjoyable if we wait, why don't we just suck it up and do it? Because that period of delay can be so uncomfortable. It involves feeling dissatisfied and unfulfilled, and sometimes we can't see beyond that discomfort.

In case you need more convincing: adding friction and learning to control your impulses has benefits beyond a more enjoyable reward. People who learn to delay gratification tend to

- make better financial decisions because they think long term

3   Chris Williamson, host, "How to Reset Your Brain's Dopamine Balance—Anna Lembke," *Modern Wisdom* podcast, episode 392, posted November 1, 2021, YouTube, https://www.youtube.com/watch?v=8l9i6-iIvYM.

- perform better in their professional lives because they have learned to strategize more effectively
- build stronger relationships because they know how to postpone their desires, making them more empathetic and patient
- have better health because they know how to endure the discomfort of working out and resist the urge to overeat[4]

Bottom line: effort is good. It can improve your overall well-being and make you nicer to be around, in addition to making the reward a little sweeter.

FRICTION-REWARD RATIO

If you're an athlete, you know there should be a balance between training and recovery. You want to push your muscles but not to the point of overtraining and injury.

Friction is similar: you need to find the right ratio between effort and reward—enough to make you work and delay gratification, but not so much that you give up

---

4   Ken D. Foster, "8 Benefits of Delayed Gratification," September 3, 2023, https://kendfoster.com/benefits-of-delayed-gratification/.

and go straight for the reward. There isn't some one-size-fits-all ratio. Finding the optimum balance is an individual thing.

If you're trying to lose weight, for example, you might insert the friction of running three miles before you can have dessert. That probably won't work for someone who hates running. Instead, they might decide to have a No Dessert in the House rule, so they have to drive to the store if they want something sweet. With social media, you might decide that you can't scroll Instagram until you've knocked out your chores for the night—dinner, dishes, taking out the trash, whatever. Someone else might need an even harder line: fifteen-minute limit and only accessed through a laptop.

We need to do the same for our kids. We don't want to raise children who get whatever they want with minimal effort or friction. The result can be a sense of entitlement and an absence of actual achievement—a Joffrey, for you *Game of Thrones* fans.

When children grow up with instant gratification (or without delayed gratification), aspects of their personality don't mature. There's no sense of accomplishment and pride in hard work. They don't learn key skills like conflict resolution and managing disappointment. Friction in childhood develops empathy and compassion, qualities we currently see atrophying in adults.

HOW TO INTRODUCE FRICTION

Friction points will differ depending on the person and the addiction. Someone who spends way too much time on Facebook checking likes and comments will probably need to set different boundaries than someone who spends way too much time gambling.

As mentioned, I deleted Instagram and X from my phone, the two apps that sucked most of my time. The friction introduced as a result was that if I wanted to use either platform, I had to log in via web browser—not a huge pain, but enough to make me consider whether I really wanted or needed to log in.

When I put Instagram back on my phone, I again set up boundaries: I limited the time I spent responding to DMs to an hour on Saturday. After a couple weeks, I adjusted it to ten minutes a day or a little over an hour a week. By setting this limit I've been able to connect with my friends and prioritize the other things I care about without slowly depleting my free time and cognitive bandwidth.

Here are a few other ways to introduce friction around smartphone usage:

- Get a Light Phone—a phone that's available for voice calls and SMS text only.

- Get the Brick app—a physical device and software that temporarily blocks certain apps and notifications from your phone.

- Log out of your apps—at the end of every day or every single time you use them—so you have to log in each time.

- If you use apps like Slack for work, delete them from your phone and only use them on your laptop or desktop.

- Make it difficult to use your phone at home by leaving it in a dedicated spot when you are home, for example, by the front door or just inside the garage—**not in the living room, kitchen, or bedroom where you'll be watching TV, eating, or sleeping**. This way, any use of social media or apps can't be done easily in a place where you unwind, connect with loved ones, or sit and think.

• Set up no-phone time zones, for example, from five o'clock, when you leave work, until eight in the morning, when you go back.

• Put a physical barrier between yourself and your device by locking it in a Mindsight timed lock box, even if it's only for a two-hour window or at specific times like during dinner.

• Change your phone's display to grayscale. iPhone users: you can even automate this setting so your phone automatically switches to grayscale at certain times of day—like right before bed. Grayscale makes your phone less appealing so you're more likely to put it away.[5]

---

5   Pranay Parab, "You Can Automatically Make Your iPhone Less Addicting at Bedtime," Lifehacker, October 9, 2023, https://lifehacker. com/use-greyscale-make-your-phone-less-addicting-1850912330.

EXERCISE 4: PICK THREE

Using the list of friction suggestions included here and/or ideas you've gathered from other sources, pick three specific actions you're going to start taking right now.

As you go along, you may need to reassess and scale up your friction. For example, with Instagram, I added the app back on my phone and set boundaries, and that worked fine. With other apps, I found I was still tempted, so I started using the Brick to restrict usage for certain periods of time. In some instances, it was still too easy to scan the Brick, so I started locking my phone in the Mindsight for periods of time. When I needed to use my phone for work, I would Brick it and throw the Brick into the lockbox for four hours. That way I could receive work calls but couldn't access certain social apps. I continually tested the friction and scaled it up based on the application and potential "need" to use it.

MY JOURNEY

Action

I added Instagram back on my phone because that's where I engage with friends and family, more than text messages in many cases. I already have in-person relationships with these individuals, or I want to maintain those connections until we meet in person. We send memes back and forth, and for me, this interaction has value.

I also made a dedicated spot for my phone on the counter by the front door. That is where it stays on a charger until I leave. If I need to make phone calls, I put Bluetooth headphones in, but the phone stays on the charger. If I need to check something on the device, I go to its dedicated spot. No more mindlessly scrolling on the couch while watching TV.

When I need uninterrupted focus time, I put my phone in airplane or focus mode and lock it in a Mindsight lockbox. I was using a Brick Bluetooth gadget for locking my phone but found it too easy to lock/unlock. The Mindsight is a physical box with a timed locking mechanism.

If I were a doctor or someone who needed access to my phone at all times for emergencies, I would probably use the Brick to disable all nonessential features on my phone and lock the Bluetooth key in the Mindsight. This way I could still receive emergency phone calls.

Another simple yet surprisingly effective way to insert friction: I purchased a matte screen protector and put my phone's display color settings to grayscale. This has the effect of making my device look like "e-ink." I immediately noticed a difference. Scrolling was way less interesting, and features like the camera were harder to use, so my usage dropped. After a week I switched back to color and was blown away at how borderline psychedelic and visually overwhelming it seemed.

Result

I am more present cognitively and more productive than I have ever been at home. I'm not going to lie; I definitely realized how addicted I was to my smartphone the first time I sat on the couch and compulsively reached for it.

Lesson

Social media, and Instagram in particular, was a form of procrastination that had become a habit. I would check it first thing in the morning, and then suddenly I had burned thirty minutes of my day. Almost every "pickup" was initiated by habit. I also realized that if I did the same thing

> *on other platforms, the time would really add up. Ounces*
> *equal pounds.*

MORE EFFORT, MORE BALANCE, BETTER LIFE

Have you ever heard of the Stanford marshmallow experiment? Back in the 1960s, psychologists did a series of tests to study delayed gratification.

Children aged three to six were offered a choice: get one marshmallow now or wait fifteen minutes and get two marshmallows. The marshmallow was placed on a plate in front of the child in the room without distractions—no toys, television, or other kids. Just the child and the marshmallow.

During that time, the children tried various tactics to keep from eating the marshmallow: sat on their hands, pulled their pigtails, covered their eyes, turned away from the plate, kicked the table, stroked the marshmallow as if it were a stuffed animal. In some ways, these actions can be seen as forms of friction or effort to delay gratification.

Researchers discovered that approximately one-third of the one hundred children waited long enough to receive the second marshmallow. They also found that age was a big factor: the older the child, the more able they were to wait.

When researchers followed up on these children years later, they made an interesting connection: children who delayed gratification tended to have better SAT scores, had attained higher education, and were overall cognitively and socially better-adjusted teenagers.

Friction doesn't just help in the moment; it can help us develop positive, socially advantageous qualities that lead to a more enjoyable life.

I'm not saying that adding friction is easy. As with any kind of growth, the process is challenging. If you're training for a marathon, the effort is physically demanding. If you're studying for the bar, it's mentally demanding. If you're moving on from a relationship, it's emotionally demanding. But in each case, the effort is necessary.

In my opinion, you are never stationary as a human. You are either growing or atrophying. You are moving forward or backward—personally, spiritually, professionally, you name it. Always strive to put the effort in, and don't let the old man in.

When tackling a task like this, it can be challenging to keep yourself on track. Most of us do better with some kind of accountability system, which we'll cover next.

PROMPT 5: MAKE IT PERSONAL

You've read my journey. You've read the ideas for adding friction. Now it's time to make it personal.

Don't underestimate the power of friction and effort in this process. We live in a world in which corporations are constantly looking for ways to introduce time savers in the name of conversion optimization for sales. You have to make a conscious effort to resist.

In Exercise 4 you were challenged to make a list of three ways to introduce friction. Now I'm asking you to reflect on why you're taking these steps and what it will look like when you do:

- Where do you think you should introduce more effort and friction in your digital life?
- Why do you think you need it there?
- How will you introduce more effort? (Remember Exercise 4.)
- What do you think the result will be?
- Do you think you can realistically maintain these friction points long term?
- Why is adding friction to this aspect of your life important to you?

QUICK FACTS

How likely are people to complete a goal?[1]

- **25 percent**: if they consciously decide to pursue a goal

- **50 percent**: if they plan how to do it

- **65 percent**: if they commit to someone

- **95 percent**: if they have a specific accountability appointment with someone (in other words, they only have a 5 percent chance of not achieving it!)

---

1   Stephen Newland, "The Power of Accountability," AFCPE, 2018, https://www.afcpe.org/news-and-publications/the-standard/2018-3/the-power-of-accountability/.

SEVEN

ACCOUNTABILITY

"Accountability is the glue that ties commitment to the results."
—BOB PROCTOR

T he various tools presented so far—journal prompts, values exercise, assessment ideas, friction sugges- tions—are meant to set you up for success in your battle against Big Tech. The stats are meant to illustrate the effectiveness of each tool in a general way—to prove that the tool really does work and can have great side benefits.

Accountability is no different. Statistically speaking, those who have some form of accountability system, be it a gym partner for fitness goals or a sponsor in AA for sobri- ety, are generally more successful at reaching their goals.

It even works for skydiving. Over a decade ago, I had one of my best friends join me in getting our skydiving licenses. It was something I needed to do for work, but my fear of heights and the risk associated with the unknown made doing so a challenge. Having a friend share in the journey and appointments for school kept me honest about following through. It wasn't until my sixth jump that things clicked and I started having fun. If not for my friend, I would have easily justified quitting. Today, I have over a thousand wingsuit flights and skydives.

Accountability doesn't need to be some formally arranged relationship. Sure, you can agree to meet Jack at the gym at five o'clock every weekday, but you can also have an informal "partner" in Darin, the guy behind the register at the Raceway gas station, who sells you those world famous tenderloin biscuits and never shies away from a good conversation. Ask Darin to bust your balls if you don't show him your Screen Time on Saturday mornings and you'll be more likely to stick with it.

As mentioned in the last chapter, inserting friction is hard. There will be times when you cross your own boundaries and bypass the effort. That's why you need accountability. We all need mentors in this process of waking up and reclaiming our minds and time.

PICK YOUR PARTNER WISELY

Accountability is a big part of addiction recovery programs like Alcoholics Anonymous. In these programs, accountability primarily refers to taking responsibility for one's actions and decisions. No more blaming other people or external factors for the addiction. Own it.

One way to take ownership in these programs is to find a sponsor—someone who has been through the steps and no longer has to use their drug of choice. This person can help the recovering addict reflect on decisions and goals, evaluate their progress, and celebrate successes. These regular check-ins provide a safety net and help prevent relapse. Plain and simple, "knowing they will be held accountable for their actions motivates individuals to stay committed to their recovery goals."[2]

The same is true when you're recovering from a digital media addiction.

I don't necessarily think you need to find someone who's struggled with smartphone overuse issues, but I do

---

2   HealingUS Communities, "The Power of Accountability in Addiction Recovery Care," February 12, 2024, https://healingus.org/blog/the-power-of-accountability-in-addiction-recovery-care-why-responsibility-is-key/.

think you need to pick your accountability partner wisely. Your phone is a private, intimate part of your life. It might be awkward to have your mom see how much time you spend swiping on Tinder.

Better to pick a trusted friend or mentor—someone who will not judge you. If you're afraid of being judged, you're probably not going to be open and honest about your usage, which defeats the purpose of accountability. You may become even more secretive, which may drive you deeper into the addictive behavior.

You also have to know yourself: Do you respond to having a drill sergeant in your face every hour of the day? Or do you respond to a more relaxed approach, where you ask the person to check in with you once or twice a week? Pick someone who matches your preferences.

EXERCISE 5: POTENTIAL PARTNERS

1. Make a list of three potential accountability partners. Whether they are friends, family members, colleagues, or teammates, they need to be trustworthy and nonjudgmental.

> 2. Give yourself a deadline for picking one of these people and approaching them about holding you accountable. The Journal Tool later in the chapter will help you come up with a more complete plan.

One of the main reasons I posted my Screen Time numbers on Instagram is for accountability. I knew my friends, as well as my unknown followers, would see the numbers and say something in person or in a comment. I didn't edit the list before I posted it, so it included all of my phone activity—dating apps and all. Even if no one liked or commented on the post, it was there for everyone to see, including my future self. Months down the road, I can look back on these posts and see that I went from forty hours a week to twenty to ten. What a great feeling!

Another bonus: the very act of seeking accountability— whether that's meeting with a friend, having someone text you, or posting online—introduces friction, which we've already seen is a good thing.

Here's a bit of irony: so far, my accountability posts have received far greater engagement than my other posts, like, 300 to 400 percent the typical engagement. People are probably pausing on the image to see all my time spent on different apps, which increases the engagement on my post, which increases the reach.

MY JOURNEY

Action

As I worked through assessment, detox, and friction, I started journaling about the whole experience. For me, this self-reflection was sufficient accountability—along with documenting and posting my usage on social media. Nobody loathes my usage numbers more than me.

That said, I also started talking to two of my friends every Saturday about my weekly consumption. One of these accountability tools—journaling or talking—would have been sufficient, but being the big data guy that I am, I wanted to split test the two experiences to see if one path offered something that might be of value to report about.

Result

Journaling provided the observation effect, which greatly reduced my consumption. Posting my numbers to social media further reinforced that effect.

Having not one but two people hold me accountable to the boundaries I established was also extremely effective. I think about my friends every single time I pick up my phone to open an app and ask myself, Is this absolutely necessary, or is this because I'm bored? The added benefit is that I get to spend more focused and meaningful time with people I care about, and we get to reinforce what it means to be friends.

Lesson

At first, it was easy for me to brush off accountability—especially related to an addiction or compulsion around behavior I see in a lot of people. I have to accept that billions of dollars are spent by the smartest people in the world and the most effective machine learning algorithms are working against me. And as the data suggested, I really shouldn't go at it alone.

PRIDE AND SHAME IN ACCOUNTABILITY

Another part of addiction programs like Alcoholics Anonymous is that people earn chips for the length of time they're off their drug of choice. Those who have gone a distance with sobriety can take pride in something that represents years, if not decades, of effort and personal accomplishment.

If they relapse, however, they have to give back the chip and start over. That threat of shame, from peers or oneself, can be enough accountability to keep people from drinking.

Pride and shame are the yin and yang of accountability, the positive and negative forces working together to keep you "sober."

Can you apply the same principle to your journey? What milestones in this process related to digital usage can you take pride in? Are you looking for complete abstinence, like people in Alcoholics Anonymous? If so, what are your milestones? Days, weeks, months, years?

Or do you measure your progress in usage—for example, the number of hours per day or week?

What is your chip? What is the consequence for relapsing?

Pull out your journal and answer these questions. When you get to the Journal Tool later in the chapter, you'll be prompted to solidify an accountability plan around these ideas.

THE POWER OF ACCOUNTABILITY

In case you missed it: you need accountability. Period. End of discussion.

Journaling can be immensely helpful, and I highly recommend it as an exercise in self-reflection and self-discovery. However, journaling is also subjective. You're recording your thoughts and feelings as well as your rationalization for why you *need* to keep playing *Candy Crush* two hours a day or why you *need* to watch every One Bite pizza review the algorithm serves you. Don't get me wrong; I enjoy a monster 8.1 from time to time. But we're trying to establish some boundaries here.

Research clearly shows that you are more likely to meet your goals and change your behavior if you have set check-ins with someone. Did you catch the Quick Facts at the beginning of the chapter? **When you schedule an accountability appointment with someone, you have a 95 percent chance of meeting that goal.**

That's the power of accountability.

Weight-loss programs like Weight Watchers make good use of this accountability factor. Back in the day, Weight Watchers required a weekly in-person weigh-in. That alone would make many people think twice about eating that candy bar. Today, those meetings happen online as well as in person, but the scheduled accountability piece still works: people who attend tend to lose two times more weight than those who do it on their own.[3]

Another big factor with Weight Watchers is the sense of community people find. They can connect with others who are going through the same process, suffering the same setbacks, and enjoying the same victories. They're surrounded by people who get it.

Community is the next tool to put in your gardening kit.

PROMPT 6: CREATE AN ACCOUNTABILITY PLAN

Use your journal to come up with an accountability action plan:

1. Make sure the person you identified in Exercise 5 is the best accountability partner— the person you trust most and who is least likely to judge you.
2. Decide how often you're going to check in with

---

3 "What Are Workshops?" WeightWatchers, accessed October 18, 2024 https://www.weightwatchers.com/us/.

them about keeping you accountable. Is it daily? Weekly? What day? What time?

3. Decide where you want to meet. Over coffee or lunch? Is it at the gym?

4. Decide what you want to be held accountable for. Is it total screen time? Is it a specific app or category?

5. Decide how you will provide proof. Is it taking a screenshot of Screen Time?

6. Is there a consequence for relapse? What is your chip to surrender?

7. How do you feel about talking to someone about this area of your life?

QUICK FACTS

The decline in male friendships, 1990 versus 2021:[1]

- **55 percent versus 27 percent**: the percentage of men who have six or more close friends

- **3 percent versus 15 percent**: the percentage of men who have zero close friends

- **45 percent versus 22 percent**: the percentage of young men who reach out to friends in tough times

---

1   Daniel A. Cox, "Men's Social Circles Are Shrinking," Survey Center on American Life, June 29, 2021, https://www. americansurveycenter.org/why-mens-social-circles-are-shrinking/.

EIGHT

COMMUNITY

"Loneliness kills. It's as powerful as smoking or alcoholism."
—ROBERT WALDINGER

W hat is the single biggest factor in most people's overall well-being?

Is it having a rockin' career? Feeling successful? Being fit and healthy?

According to Steven Crane, a Stanford researcher, it's our relationships. Having a healthy network of social connection "constitutes the largest single factor in the overall well-being of most people."[2]

2   Carly Smith, "How Social Connection Supports Longevity,"...

That said, loneliness in America has gotten so bad that US Surgeon General Vivek Murthy called it an epidemic with **health risks on the same level as smoking fifteen cigarettes a day.** Murthy's report concluded that **"loneliness can increase the risk of premature death by 26 percent and raise the likelihood of heart disease, stroke, anxiety, depression, and dementia."**[3]

This makes sense when you consider we are social creatures. We have lived in community with others for as long as we have been on the planet. We developed connections because they allowed us to scale efforts and divide tasks for the benefit of all: some could forage; others could gather water; others could hunt.

There's another reason for our motivation to connect: our brain releases dopamine when we do, and as we've already discussed, that dopamine hit feels good. We also get a dopamine release related to all of the meta experiences around community: when people like us, agree

---

…Stanford, December 18, 2023, https://longevity.stanford.edu/lifestyle/2023/12/18/how-social-connection-supports-longevity/.

3 Margaret Osborne, "An 'Epidemic' of Loneliness Threatens Health of Americans, Surgeon General Says," *Smithsonian Magazine*, May 10, 2023, https://www.smithsonianmag.com/smart-news/an-epidemic-of-loneliness-threatenes-health-of-americans-surgeon-general-says-180982142/.

with us, or enhance our reputation; when we experience the same emotion at the same time as another person.

Think about social media: it's a virtual frenzy of community-related dopamine spikes: we have endless opportunities for likes, affirmative comments, and connecting with people who are outraged at the same event or post.

Are you a fan of watching reaction videos? Ever wonder why they're so enjoyable? Dopamine is released when we watch someone watch a video and react the same way we do. Combine that with a social platform that creates an algorithm whose purpose is to keep you watching longer, and this type of content gets pushed more often.

There's a problem, of course. Social media has taken it way too far. It has drugified human connection because it provides all of the factors that enable addiction: access, quantity, potency, and novelty.

In olden times, you actually had to get off the couch, go out, and meet people. Now you can sit there in your boxers and swipe right or left. You also have an endless supply of reaction videos and posts by people expressing emotions you identify with. The connections we feel are heightened because social media combines several "drugs"—beautiful images, pleasing music, sex, flashing lights—to increase the potency of the dopamine release.

And machine learning algorithms know just what we will respond to. Dopamine is sensitive to novelty, so these systems have figured out how to identify what we like and offer a new, slightly modified version.[4]

If we remove or at least limit that huge source of dopamine in the form of social media, we need to replace it with the real thing. We were made for connection, and the feelings of pleasure associated with it are real and valid. We simply need to start getting reinforcement through true, in-person interactions, not the imposter found online.

The ideas we'll talk about in this chapter are not revolutionary, but they're a needed reminder as you disentangle yourself from the often isolating influence of Big Tech and digital media.

ANTISOCIAL MEDIA

I'm going to say it again: we are social creatures. We are made for connection. But social media has lured us into a poor and highly addictive substitute. Ironically, this substitute has actually made us more isolated. It didn't really bring us together as promised. Now we're all sitting at home alone

---

4   Williamson, "How to Reset Your Brain's Dopamine Balance.".

on our phones, "connecting" with others. We're becoming less social, less empathetic, and more disconnected.

Even when we are together, we are increasingly distracted and distant because we're on our electronic devices instead of interacting with the people in front of us. Our time on social media cuts out a crucial piece of building community: actually being with that community, interacting in person around shared interests.

The original value proposition was being able to connect with people at scale. But once again, we have incentive misalignment throwing a wrench into the situation: the social media platforms don't really want to help you stay connected with friends and family across the globe. They want your attention. They want to keep you on platform so your behavior can be validated so you can be offered more targeted advertising so they can make more money.

We humans want to be liked. We get a dopamine boost when we think we are understood and appreciated. Social media has hijacked this natural response and turned it into a way to make money. Every time you click that Like button, you give Big Tech an opportunity to validate your behavior, show you something new and exciting, and keep you coming back, which gives them a chance to show you more targeted ads and make more money.

Another problem with social media is the parasocial component—a one-sided connection between a user and someone they don't know. Decades ago, the "someone" was likely a celebrity who seldom broke the fourth wall in film and TV. Now the line has been blurred as creators talk directly to the camera and users can comment and engage digitally, giving the viewer a feeling of connection even if the communication isn't reciprocated. On a more serious level, some users can start substituting experiences with others on digital platforms, from Instagram to OnlyFans, for in-person interactions and can neglect, strain, or never develop the relationship with a partner.

Beyond isolation, social media has contributed to the bifurcation of society. With the sensationalist thumbnails and headlines, it is optimized to divide. I'm not saying Mark Zuckerberg is trying to make everyone angry. The algorithm simply learns what gets the most engagement, and that's the content that launches people into a fight-or-flight frenzy.

Online, it's easy to hide behind your keyboard and blast out reactions. In person, with people you actually know, you might be less abrasive and more careful with your words. That's a good thing.

BENEFITS OF COMMUNITY

So you've decided to limit social media and start getting your connection dopamine injection in person. Now what?

Considering your goal here, I recommend finding a community around an activity that must be performed without your phone. CrossFit, pickleball, running club, SoulCycle, archery, board game meetups—you get the idea. The goal here is to make connections, talk to people, find common interests, have engaging conversations, laugh, struggle, and/or grow together. As previously established, you can't do that when you're glued to your phone.

You might be tempted to take photos or video of the activity to share on social media. Don't. You're doing this for the benefit of the activity, not the benefit of others' perception of you.

EXERCISE 6: FIND YOUR TRIBE

1. Make a list of at least five activities, hobbies, clubs, or groups you can join to build community.

> 2. Give yourself a deadline: when are you going to start getting involved?

In your community-seeking endeavors, you're there to find common ground, not to judge. Who cares what your SoulCycle buddies think about the latest partisan bill introduced or geopolitical conflict in a foreign region somewhere? You're trying to let the brain reset, remember? Don't jump into discussions that get your fight-or-flight response all hyped up again.

You can reap even greater rewards from community when you do hard things together. Take Orangetheory, for example, the boutique fitness studio that offers intense one-hour whole-body classes. One study shows that Orangetheory has a retention rate of over 80 percent. Compare that to the retention rates for most globo gyms, which hovers around 60 to 75 percent.[5]

Why does Orangetheory have a higher retention rate? My guess is that people crave the comradeship that comes with being part of a smaller, close-knit group that

5   Henry Sheykin, "7 Critical KPIs for Orangetheory Fitness Franchise," FinModelsLab, October 29, 2024, https://finmodelslab.com/blogs/kpi-metrics/orangetheory-fitness-franchise-kpi-metrics.

kicks butt together. You just don't get the same connection and community when you're doing your own thing at a big gym.

It's challenging to be a white belt among the black belts, to be the new guy on the team, or to be constantly corrected by the piano teacher because you keep hitting the wrong key. But like the cold plunge, enduring the hard part leads to great rewards—the long tail of the dopamine release. When you tackle something new and challenging with others, you get multiple benefits: new skills, improved health, good times, and new friends.

One warning: it's easy to trade one addiction for another. If you've been spending forty-plus hours online and now you're not, you might be tempted to put all of those hours into CrossFit or horseback riding. Go back to your values list. What were you giving up because you were so isolated online? What actions can you take so that your time is more in line with your priorities?

Optimize the give-and-take so that you spend your time on the things that matter most. That's the algorithm to live by.

MY JOURNEY

Action

During the process of writing this book and testing different tools, I've had to assess relationships with people as well as digital products. I had already begun establishing my core values and boundaries. When I got to the community component, I had to reflect hard on what I was neglecting; where I wasn't getting or receiving value and, more importantly, why; and where I needed to make adjustments. In the interest of building a truly supportive community, I had to start cutting certain people out of my life—some who had been a big part of it. Our values and paths were not in alignment.

Result

The whole process of evaluating and then taking action was extremely difficult. In the end, however, I was almost immediately relieved after having the conversations. I had been coasting through so many aspects of life as a passenger, not really in control of where I was going because I was letting people steer me down a path I didn't choose and that didn't make me happy because we didn't share a vision or values. By parting ways with them, I opened up my life and schedule in a way that has already facilitated finding new people, communities, and tribes that bring me joy and fulfillment.

Lesson

Some friction is hard. Real hard. Especially when it comes to community and who we share our journey with. Humans have evolved as social creatures. It's ingrained in our DNA to belong to a collective greater than any one individual. It's paramount to carefully choose our community and those we share the journey with.

CONNECT IN YOUR VALUES

Isolation isn't healthy. We humans were made for connection and community—the real in-person kind, not the kind manufactured by unsocial media.

When you're looking for community, look for opportunities for connection rooted in your values: family game night, guys' weekend fishing, hiking club. Use that valuable commodity of time in a way that supports your digital disengagement journey *and* your priorities. Just don't go too hard in the paint in any one direction. Balance is the key.

In an effort to activate the community around digital awareness and protection of our hearts and minds, I offer some freebies on my website (*www.WarriorsGarden. com*). Keep in mind that these are tools, and no single tool is meant to do every job. Everyone's journey is different. Some tools may help; others may not. You may even find different tools work. If you do, please share with the rest of the community so we can all rise together.

PROMPT 7: STAYING BALANCED

Staying balanced in any area of life requires constant self-reflection.

Take time to look back at the different moments in your life where you felt part of a community. In your journal, write about who you were, what your life was like, who those people were, why those moments were significant, and how they made you feel. You can do this once, once a day for a week, once a week or month, or indefinitely.

Here are two examples to show you what I mean:

One easy example of when I felt part of a community was when I played basketball as a child. I remember the excitement of the games on Saturday afternoons. The sound of squeaking shoes on the court. The parents yelling in the bleachers. The thickness of DIY Gatorade mixed at the wrong ratio. The after-game all-you-can-eat buffets at Pizza Hut. We put in the work during practices. We shared in the joy of a win and the frustration of a loss. It was fun, it was exciting, and it was great to be part of a community and team of friends in that shared experience. We belonged.

Another example is the sense of community I felt on September 11, 2001. Like everyone else who is old enough to remember, I know exactly where I was when I heard what happened. I was getting ready for work. I had a specific process for making Pillsbury Toaster Strudels. I'd put them in the toaster on a low setting and then take a shower and get dressed while the strudels defrosted. I'd cut on the TV to the news and push down the pastries one last time to get them crispy on the outside and hot and gooey on the inside.

When I sat down to eat, I saw a plane had crashed into one of the World Trade Center towers. *How could that have happened?* As I started to take another bite, I saw the second plane fly into the other building. I was confused. I was scared. *Do I go to work? Are we at war? What is going on?*

Through the chaos that ensued over the next twenty-four hours, I found an intense sense of belonging to my community. Everyone called each other to check in. "Are you okay?" "Do you need help with anything?" "Can I get you anything?" From family, friends, and strangers on the street, I felt love and gratitude that we were all on this speck of molten lava dust together, hurling through space and time at 140 miles per second.

Your turn. When did you feel part of a community? Describe everything you remember about the people, your feelings, and your surroundings.

QUICK FACTS

Negativity is bad for you.[1] It can

- weaken your immune system
- make you more prone to a heart attack or stroke
- affect your intelligence and ability to think
- impact reasoning and memory

Gratitude is good for you.[2] It can

- enhance your relationships with deeper connection and more intimacy
- increase your optimism and general satisfaction with life
- increase your self-confidence and self-esteem
- improve sleep
- decrease your experience of frustration, envy, and regret
- decrease your likelihood of experiencing depression and PTSD

1   Elle Kaplan, "Why Negative People Are Literally Killing You (and How to Protect Your Positivity)," Medium, November 14, 2016, https://medium.com/the-mission/why-negative-people-are-literally-killing-you-and-how-to-obliterate-pessimism-from-your-life-eb85fadced87.
2   Courtney Allison, "How Gratitude Is Good for Your Health," *New York Presbyterian*, November 16, 2023, https://healthmatters.nyp.org/is-gratitude-good-for-your-health/.

NINE
GRATITUDE

"When we focus on our gratitude, the tide of disappointment goes out and the tide of love rushes in."

—KRISTIN ARMSTRONG

On December 5, 1914, Ernest Shackleton and his crew of twenty-eight men, sixty-nine dogs, and one tomcat left South Georgia for Antarctica. Their goal: establish a base on the Weddell Sea coast.

If you've heard the story, then you know they never made it.

Two days after departure, their ship, the *Endurance*, entered the thick barrier of sea ice that surrounds Antarctica. About a month later, the *Endurance* got trapped in the pack ice. One crew member said the ship was "frozen like an almond in the middle of a chocolate bar."[3]

There was nothing they could do but wait out the winter on board—and winter in Antarctica is ridiculously long and cold. On May 1 the sun vanished and didn't reappear for four months. On June 22, despite the harsh circumstances, they celebrated Midwinter's Day with a feast.

Then the pressure from the ice started crushing the *Endurance*. On October 27, Shackleton ordered everyone to abandon the ship, and within a month, the ship sank.

The crew spent the next six months living on the ice floe and using small boats to row toward land, which they finally reached on April 15, 1916—497 days after they started the journey. It was another four months until the last of them made it back safely to South Georgia.

How do people survive such grueling circumstances? How do they maintain a positive outlook that enables them to keep trying despite exhaustion, hunger, sleep deprivation, and intense cold?

Perhaps more importantly for us, how do you become the kind of person who is an asset during hard times,

3   Kieran Mulvaney, "The Sunning Survival Story of Ernest Shackleton and His Endurance Crew," History, October 21, 2020, updated May 2, 2024, https://www.history.com/news/shackleton-endurance-survival.

someone other people want on the boat because of their positive can-do attitude?

Not by consuming a lot of digital media.

When you're constantly in a hyped, agitated fight-or-flight state, you're like a caged wild animal, always on guard. As opposed to a wild animal that just ate and is now chilling on the savanna, calm and at peace, enjoying true community with the pride.

Plus, social media is filled with mind-numbing, soul-sapping negativity. How can you maintain a positive outlook in tough times when you're weighed down with hopelessness?

You can't. To be the person others want with them during challenges, you need to guard your mind. For me, the best way to do that is by developing a gratitude practice.

The point isn't to gamify gratitude so you earn acceptance. Studies around gratitude show it results in feeling happier, and I like to think people want to feel happy over the opposite. Gratitude is a way to course-correct from the negativity and polarization that social media inherently feeds us as a by-product of its need to extract the value it seeks, namely our attention.

In this chapter we'll discuss tools like self-reflection and other gratitude practices to replace the negativity and become someone people want to be around.

COMBAT NEGATIVITY

If you spend excessive amounts of time online, chances are good that your general outlook is being hijacked by negativity. Thanks to our innate "negativity bias," we humans tend to give more weight to negative information or events than to positive ones. This is probably another leftover from our primitive days when we needed to know what potential threats were out there—think saber-toothed tiger—so we could avoid them.[4] Big Tech knows this and takes advantage of it, so our media feeds are filled with outrage, drama, and bad news.

In this state of mind, we often neglect gratitude, though it is the very thing that's needed to realign our perspective on life and our surroundings. We may be in a state of world war dynamics and civilization may be collapsing, but we still have the ability to skew our perspective back into a healthy balance by focusing on the good in our lives.

Are you going to be able to single-handedly course-correct the situation in Ukraine or Israel? Can you alone fix the US debt? Probably not. Most of us aren't the Elon

---

4   Beeston, *Brainhacked*, 38.

Musks of the world who actually have the financial and social influence to make large-scale changes.

Don't misunderstand me. I'm not saying you shouldn't care about global affairs or try to make a difference in the world. But you also shouldn't wallow in the negativity of it all. Being constantly inundated by the clickbait thumbnails, sensationalist headlines, and fear-inducing messaging comes at a cost: your mental and emotional health.

The nonstop negativity feed also keeps you from acting on matters you can have a direct impact on—your family, neighbors, and local community. I firmly believe a huge portion of society's problems could be improved or fixed if we started focusing on the things we can control instead of stressing out about the things we can't.

Maybe you're the one person on planet Earth whose feed isn't filled with doom-and-gloom messages. You could still be adding to your negative outlook by compulsively scrolling others' feeds. Constantly comparing your normal life to others' sensational highlight reels can lead to a number of unrealistic expectations: *Why don't we get to go on vacation like them? Why don't I have that new thing? Why don't I get to have that experience? Why don't you love me like that?* Focusing on what you don't have can negatively impact your mental and emotional health. You start believing that your own life and work are way less interesting and that what you have and do are never enough.

You have to fight back. Focus on what you can control—your mindset. Big Tech certainly isn't going to help. They're making money by keeping you in an anxious, depressed state.

What if what you have right now is enough? What if you can amplify things you already have and increase your quality of life and happiness? When the world is increasingly dark, it takes concerted effort to remind yourself of the things that you appreciate and love—things you already have. But it's worth the effort. You can combat negativity by consciously cultivating a practice of gratitude.

WAYS TO CULTIVATE GRATITUDE

If this is all new to you, start small. For example, start saying thank-you to the people in your life: coworkers, friends, significant others, parents, children, and so on.

Get specific. Tell the barista at Starbucks, "Thank you! I am grateful for you and everything you do today. I really needed this caffeinated pick-me-up!"

Tell the person making tortillas in the bakery at the grocery store, "You are my favorite person! You are the reason H-E-B is my favorite grocery store and Taco Tuesday nights at my house are a fan favorite. I'm truly grateful for you and everything you do!"

There are so many people in your community that play a vital role in making it better, and everyone likes to feel appreciated. Seeing a smile on their face will probably put a smile on yours.

Here are a few more suggestions for practicing gratitude. Find the ones that work best for you:

- Give thanks for your food before you eat. Acknowledge a deity in the process, or just express gratitude for the food itself, the fact that you have it, and the fact that you can taste and enjoy it.

- When you get into your car, take a deep breath. Express gratitude that you have reliable transportation and money to put gas in your car.

- At the end of each day, take out your journal and write down something, just one thing, that went well that day. Or try doing this in the morning, and write about one thing that went well the day before.

- A variation on the above: Pick a time each week to write out five blessings. Be specific. Describe the sensations you felt when something good happened to you.

- When you find yourself focusing on something negative, pause and think about one positive aspect of the situation. If you really think about it (remember: use it or lose it), you'll probably be able to think of more than one.

- Tell someone why you are thankful for them. Yes, you can send a text, but there's something about the act of physically writing a note that makes a bigger imprint on our own brains. Plus, who doesn't love getting mail that's not a bill? Give yourself a goal of sending one gratitude letter a month. Bonus: writing thank-you notes is a great way to strengthen your community.

Focusing on the positive doesn't mean living with your head in the sand. You're not being naive or denying that this world has issues. Practicing gratefulness allows you to see the good you experience every day despite all the negativity and chaos. It's about balance—tilting the scales back in response to the impacts of digital.

The end result? You'll feel better. Being grateful will actually make you a happier human, which will make you far more pleasant to be around.

EXERCISE 7: START PRACTICING

1. Using the list of suggestions as a starting point, make a list of three specific ways you will incorporate gratitude into your daily routine.

2. Give yourself a deadline: when will you make this list, and when will you start incorporating your ideas?

BENEFITS OF GRATITUDE

Some of you might still be skeptical. Can gratefulness really make you happier and improve your overall well-being? According to many studies, yes. Gratitude is a crucial tool in your quest to protect your mind and soul against Big Tech.

Dr. Robert Emmons of the University of California, Davis, and Dr. Michael McCullough of the University of Miami have done a bunch of research on gratitude. In one study, they divided the participants into three groups and asked each group to write a few sentences on different topics. Group 1 wrote about things that had happened during the week that they were grateful for. Group 2 wrote about daily irritations or things that upset them. Group 3 wrote about events that had affected them, with no direction on whether they were positive or negative. Guess who felt more optimistic at the end of ten weeks? Group 1, the people who wrote about gratitude. Bonus: they also exercised more and visited the doctor less than those who focused on the negative.[5]

In another study, Dr. Martin Seligman, a psychologist at the University of Pennsylvania, tested the impact of various positive psychology interventions. For one assignment, participants wrote a letter of gratitude to someone they had never properly thanked for his or her kindness. The result? The letter writers immediately

---

5 Harvard Health Publishing, "Giving Thanks Can Make You Happier," August 14, 2021, https://www.health.harvard.edu/healthbeat/giving-thanks-can-make-you-happier.

showed a huge increase in happiness scores. The act of expressing gratitude had a longer-lasting impact than any of the other interventions.[6]

UCLA Health also wrote an article about the benefits associated with gratitude. One of the most important findings: you'll get the biggest health benefits when gratitude becomes habitual—something that is ingrained in your thought process because you practice it daily.[7]

Remember the negative impact Big Tech and smartphone use can have on mental health? Gratitude can fight that. A review of seventy studies (twenty-six thousand people) found an association between higher levels of gratitude and lower levels of depression. It seems like gratitude can lessen symptoms of depression (dissatisfaction with life and low self-esteem, for example), which allows people to see all the good things they have.[8]

In addition, gratitude can help us cope with anxiety. When we feel anxious, we're generally worrying about the past or future. Gratitude keeps the mind focused on the present, on the good things that are true right now.

---

6   Harvard Health Publishing, "Giving Thanks Can Make You Happier."
7   UCLA Health, "Health Benefits of Gratitude," March 22, 2023, https://www.uclahealth.org/news/article/health-benefits-gratitude.
8   UCLA Health, "Health Benefits of Gratitude."

Practicing gratitude has also been found to relieve stress, improve sleep, and reduce the risk of heart disease. It's good for your mind and body.

OPEN TO TRUTH

Everyone likes to think they're objective, but the fact is, we're all susceptible to confirmation bias—especially if we're stuck in a negative feedback loop. If we repeatedly hear only one side of the story, then whenever we hear the opposite side, it will seem false. **That's why it's important to remain open to the fact that you might be wrong or that the media stream you consume might not be 100 percent true.**

Gratitude can provide a counterbalance that allows you to step out of these loops and assess the amplifications. It allows you to remain grounded in the present, aware of other perspectives, and open to the truth.

Typically, raw data is objective; data sets are subjective. Show me a data set—a collection of tables/numbers—and I'll show you a perspective. Data sets are really groups of information meant to present a narrative or tell a story. In most instances the media will feed us data sets, not objective data. Opinion journalism uses confirmation bias with a known demographic, their audience, to elicit an emotional response in an effort to increase watch time and views on their show so they can sell more ads.

There are tons of real-world examples, but I'll pick one: skydiving statistics. Earlier I mentioned that I had a friend get his skydiving license with me because I was dragging my feet due to my fear of heights. I thought it would help to have some accountability. There's more to that story.

During my consulting years, Relativity Media asked me to pitch video ideas to promote the film *Act of Valor*. Being that it was a film about Navy SEALs, I offered a series of videos breaking down action-packed aspects of the film, one of which included HALO, a form of skydiving where a jumper exits a plane at thirty thousand feet and doesn't open their parachute until very low to the ground. Relativity Media said yes to all of my ideas, which meant I had to put my money where my mouth was and learn to skydive.

I started looking into the data on the safety of skydiving and noticed that many of the media articles I read conflated certain statistics to make the death toll look higher than it was. Clearly the writers were trying to elicit a certain emotional response. They grouped together every person who died jumping with a parachute—beginners, jumpers (people who jump off objects closer to the ground), swoopers (people who fly very small canopies extremely fast toward the ground and pull up at the last second in judged competitions)—so it was difficult to find facts on the kind of jumping I would actually be doing.

So I checked the United States Parachute Association and learned that in the last twenty years, *the number of skydives has roughly doubled, but the number of fatalities has been cut in half.* That meant skydiving was only one-quarter as risky as it was two decades ago.

Even though I was still terrified of heights, I saw how my fear was being compounded by sensational articles and media reports, so with the help of my accountability buddy, I decided to go for it.

There's no shortage of hot-button topics on social media that feed on our innate negativity bias and generate heated debates through selective data sets. Where do you stand on *Roe v. Wade?* Where do you stand on firearms? When was the last time you questioned the data or your logic around a conversation that emotionally charges you?

I'm not trying to sway your opinion on any topic. The point here is to illustrate how easily we can be deceived because we're stuck in a negative feedback loop hearing the same (possibly wrong) point of view over and over. I also want to give you yet another reason to cultivate gratitude. If you can remain open and receptive to truth, even if what you hear doesn't match the data and arguments you've been fed, then you're more likely to make sound decisions.

FOR THE PARENTS

Whether we're talking about adding friction or building community or developing a gratitude practice, these things can be learned from a young age.

If you have kids, start now. Help them see the value in community-building activities, whether they join a sports team, band, or 4-H club. Help them develop a gratitude practice so they have an ingrained appreciation for the little things and don't grow up with a sense of entitlement.

Open to Empathy

Along with gratitude, empathy makes you someone people want on the boat. Unfortunately, that social-emotional skill is atrophying big-time in our society, possibly because we have an epidemic of main character syndrome. Everyone thinks they are the protagonist in their own life story as well as everyone else's, leaving little room for thinking about others.

As main characters, people post highlight reels of their life, from new cars to vacations to highly edited photos with perfect lighting. Followers who consume this content, especially young people, can develop unrealistic expectations and consciously or subconsciously see this flashy, picture-perfect life as the norm.

They also post updates on every little detail of their lives—the clothes they're wearing because of the weather, the face they make because of a trending topic, the twelve photos of their lunch. Don't get me wrong; I love a good Scratch Sushi experience, but as a society we're becoming so wrapped up in our own stories that we lose the plot of our shared human experience. I believe this narcissism is contributing to a widespread lack of empathy, which is probably another contributing factor in our societal polarization around politics in particular. We cannot get outside of our own heads to consider the experiences and perspectives of others.

But it's even worse than that. Our societal lack of empathy is causing us to view people who disagree as inherent threats, not just people with different perspectives or opinions.

Gratitude can help us change that. It can shake us out of our main character syndrome and help us become aware of people and situations around us. As we become more aware, we learn to appreciate all that we have. Gratitude opens the door to empathy toward the people in our lives, which allows us to connect with our world in a deeper and more meaningful way.

MY JOURNEY

Action

I purchased a handful of gratitude journals to help me along in this process. I found a guided approach with steps to be the most useful.

The first step was noticing or observing what went on in my day. In my journal I documented answers to questions like "What happened today?" "Did something good happen to me?" "What can I be grateful for today?"

After a week or so, I moved on to reflecting and appreciating. In my journal I added prompts like "What brings me joy, and would I be sad if it were missing from my life?"

After another week I moved on to practicing gratitude daily in a number of ways. My journal prompts now are "What am I grateful for today?" "What am I grateful for in my life?" And I still add once a week "What would I be sad about no longer having in my life?"

In addition to the journal prompting for gratitude, I have implemented conscious appreciation in the moment. For example, before meals I make a concerted effort to be grateful for the chain of people that led the food to my plate, the time in which we live and the abundance we enjoy, and the nourishment that the food will provide my body so it can do the things I ask of it.

When I get into my truck and it starts up, I am grateful for reliable transportation. I am grateful I have a career that has facilitated the financial ability to afford my truck. I am grateful for the many, many people who helped make the components, from the fuel to the roads, so I can go from my home to where I need to be.

Another way I incorporate gratitude is through my breath work and meditation exercises. My acting coach, Jeff Marcus, introduced me to guided meditation in my early days as an actor. Over the years I've found this practice to be extremely beneficial. In recent years, many scientists and researchers like Andrew Huberman have touted the benefits of non-sleep deep rest, or NSDR. I do a combination of a chakra meditation with NSDR and gratitude practice. It subjectively makes me happier and more cognitively present, and I have objectively documented that it substantially lowers my blood pressure and resting heart rate.

Sound interesting? I have outlined steps and curated playlists from others at www.WarriorsGarden.com.

Result

Some people might think it's absurd to thank construction workers before they go for a drive, but as I did this, something truly wonderful started to happen. I started consciously acknowledging so many aspects of my life I was taking for granted. I realized that those guys working on the curb were part of my community. I appreciated them. They were an integral part of my life experience. All of a sudden my community grew and grew. I felt happier. I was grateful. And even though I know none of these people, I know I'm not alone. We are all doing this life together.

Lesson

Going through the journaling exercises and then moving into moment-by-moment appreciation made me realize I could practice gratitude in a way that didn't take too much

effort—before a meal, before a drive, and as I journaled. My
results corroborate what the scientific literature has shown:
I am happier.

BE THE PERSON YOU'D WANT ON THE BOAT

In *Meditations* 8.47, Marcus Aurelius says that we shouldn't stress about any one thing unless it affects us physically, since the severity of the problem depends on our perception of it. Given that, we have the ability to make any problem seem minor.

Meaning: unless someone is stabbing you with a knife, it's within your ability to make the psychological change that impacts your life in a positive way. And one of the best tools for doing that is gratitude.

In life, one could argue, character is way more important than skill. Someone could be an amazing developer or welder or whatever, but if he is a miserable person and no one wants to be around him, then the whole organization or team suffers as a result. At that point, you gotta cut the cancer out. It doesn't matter how skilled that person is. It's not worth the mental and emotional cost to everyone else.

On the other hand, some people are force multipliers. They may not have the strongest skillset or be the most qualified, but they know how to boost morale and rally the troops when challenges arise. They can bring the community together so they don't fall apart. These people solidify the team and make it better.

Be the force multiplier, the person you'd want on the boat with you. Focus on the positive and incorporate a consistent practice of gratitude for the little things and the big.

PROMPT 8: GRATITUDE FOR THE LITTLE THINGS

A couple of reflection exercises related to gratitude:

1. Write about a moment in your life when you were in need and someone reached out to help. It could be as trivial as someone ahead of you in the drive-through line buying your coffee (caffeine can be a need) or something more serious like someone stopping to help you change a flat tire in the pouring rain on a deserted road. Write about the gratitude you felt in the moment.

2. Write answers to the following questions: "What am I grateful for today?" and "Who am I grateful for today?" Then write about these questions every day.

Is it possible to find that level of appreciation and thankfulness in the little daily things you take for granted—health, work, food, reliable transportation? It is if you consciously work at this habit.

CONCLUSION

"Life's a garden: Dig it."
—JOE DIRT

This book is not my way of passing judgment on how you're living your life. I'm simply pointing out that we're living in a world of exponential digital growth, and Big Tech is taking full advantage of it. If we continue without checking our digital consumption and its impact, we might find that certain downstream effects are almost impossible to come back from. We need to start thinking about how to course-correct before it's too late.

I'm also not trying to devalue or bastardize the value proposition of digital media or technology. Both have objectively improved quality of life, efficiencies in business, and much more. But some aspects also have misaligned incentives, none more prevalent than in the world of social media. We need to slow down and look at the possible second- and third-order consequences that result when incentives are misaligned. If we don't pause to assess what we value and what brings us fulfillment, we're going to be manipulated into a future that could become dystopian relatively quickly.

Remember, ounces equal pounds. If we individually start making changes to our personal digital consumption—if we, as modern-day warriors, protect the garden of our minds—then we can start taking care of the people around us. The world will become a better place, with more empathy, connection, and meaning, if we focus on the good we can do within ourselves and our immediate communities.

Fifteen years ago, I was oblivious to digital media problems. I contributed to and benefited from the Big Tech system. Now I see so clearly what's going on. I'm trying to practice what I preach and offer solutions to benefit myself, my family, my community, and people like you.

THE MISSION

My goals in writing this book are fairly simple:

- I want to show you, beyond a shadow of a doubt, how Big Tech is manipulating you, extracting value, and negatively impacting your life.

- I want to emphasize the disastrous impact this manipulation is having on our mental health and relationships.

- And perhaps most importantly, I want to offer tools to help you course-correct from digital exploitation—tools related to insight, detox, friction, accountability, community, and gratitude.

I don't care so much about being a Paul Revere, warning of impending doom. I want to be part of what happens after. My hope is that you add to the toolbox because I want to benefit as well. Go to my website and share what's worked for you. Give back. Help others. Contribute to the community of people striving to disconnect, guard their minds, and join the resistance against the manipulation and control of Big Tech.

THE TOOLBOX

A warrior masters their skills and keeps their tools handy. So here's an abbreviated list of the exercises and prompts from each chapter; feel free to look back for the full explanation. Experiment, figure out which ones are most effective for you, use them often, and talk to others about what you've learned so we can all improve on these tools for facing this ever-changing battlefield.

Chapter 3: Insight

- Prompt 1: Know Yourself. *Answer these questions to prime the self-reflection pump.*
- Exercise 1: Are You Addicted? *Assess whether you subconsciously and/or compulsively rely on your phone.*
- Prompt 2: Screen Time Reality Check. *Use your Screen Time numbers to reflect on the consequences of your time usage and how you might use that time differently.*

Chapter 4: Define

- Exercise 2: Values and Priorities. *Write out your core values and priorities.*
- Prompt 3: Letters to Yourself. *Write a letter to fifteen-year-old you and then to ninety-two-year-old you.*

Chapter 5: Detox

- Exercise 3: What and How. *Decide what aspects of your digital life you need to detox from and decide how you're going to do that.*
- Prompt 4: Doing Hard Things. *Reflect on the last time you did something truly difficult—why it was so hard, how long it took you, and how you felt when you finished.*

Chapter 6: Friction

- Exercise 4: Pick Three. *Pick three specific friction actions you're going to start using right now.*
- Prompt 5: Make It Personal. *Think about why you need friction in the areas you picked, what the result will be, and why taking this step is important.*

Chapter 7: Accountability

- Exercise 5: Potential Partners. *List three potential accountability partners and give yourself a deadline for picking one and approaching them.*
- Prompt 6: Create an Accountability Plan. *Decide how often you'll meet with your accountability partner, what you'll be held accountable for, how you'll provide proof, and how you feel about this whole accountability thing.*

Chapter 8: Community

- Exercise 6: Find Your Tribe. *List five activities, hobbies, clubs, or groups you're going to explore and when you're going to get involved.*
- Prompt 7: Staying Balanced. *Write about times when you felt part of a community: who you were, what your life was like, why these moments were significant, and how you felt.*

Chapter 9: Gratitude

- Exercise 7: Start Practicing. *List three ways you will incorporate gratitude into your daily routine and state when you will start.*

- Prompt 8: Gratitude for the Little Things. *Write about a time when someone helped you and how you felt in the moment.*

THE CALL TO ACTION

Okay, you've finished the book. Now what? Let me suggest three steps for starters. If you haven't bought a journal yet, you can use the Thirty-Day Challenge at the back of the book to record your thoughts.

1. Assess

Conduct an *honest* assessment of how tech is negatively impacting you. Don't remember what I'm talking about? Go back to Chapter 3 for a little refresh. Here are some questions to consider:

- Do the digital media outlets where you are consuming most really provide value, or do they steal time away from the things you value?

- How much time are they taking? What are they taking that time away from?

- Do you feel obligated to respond to texts and phone calls quickly, or can your phone be off or away from you for extended periods of time?

- If you feel obligated, is that because others have put that expectation on you, or is the need purely self-imposed?

2. Detox, Add Friction, Build Community

Every warrior has a different garden that receives varying amounts of rain and sun. In agriculture, even neighboring farms can have different soil and require different amendments.

The point? You need to decide what path is best for you. Detox is not optional, but you need to evaluate what you need: a full thirty days off of all social media? Two weeks off of certain apps? Absolutely no porn? What do you need to do?

Next, figure out what friction and boundaries you need to add.

- Do you need time constraints? For example, no social media from eight in the morning till eight at night. Instagram only on Saturdays.

- Do you need physical boundaries? For example, no phones at the dinner table. Phones have to be kept in a dedicated place when at home. Phones get Bricked or put in Mindsight during family time. Phones stay at home on grocery store trips.

Get creative, but when you set these boundaries, remember: these are the hills you die on.

Along the way, start building your community in the real world. Look for things that help reinforce your core values and principles.

- If fitness and health are important, look for a local CrossFit, yoga studio, BJJ, running club, or anything where groups get together for a fitness activity.

- If mentoring is important, look for local Boys & Girls Clubs as well as elderly, special needs, and friendship clubs. Use social media to your advantage and search for these in your local area.

- If charity is important, look for organizations around an interest you have. There is no shortage of animal rescues that need help with mobile pet adoptions and other services. There are community kitchens, construction and building programs, and so many other opportunities to have a positive impact in your community while getting to know people.

- If fun and entertainment are important, find local game clubs around your interests, whether that's cards, chess, RPG, bowling, or Ultimate Frisbee.

As you close doors in life, you will find opportunities to open new ones. Mine have developed around my core values (love, integrity, enthusiasm), specifically in the areas of fitness and volunteering.

When I first set out on this endeavor, I only saw how isolated and divided civilization was. Now I see this was a perception distorted through the lens of digital manipulation. I believe most people are kind and good-hearted. Through community you will reinforce the fabric of what makes our species thrive. United we stand, divided we fall.

3. Cultivate Gratitude and Optimism About the Future

Birth rates are decreasing and suicide is increasing because people are losing hope about the future. The narrowing of perspectives and perceptions has led to a lot of cynicism, which on a collective basis isn't healthy for society. Be the one to bring the positive yang to a negative yin: develop gratitude in your own heart and mind and spread it to others. Be the person others want in their boat.

The soil in your garden might not be ideal. So what? Take that as a personal challenge to do something about it and make it workable. Be a warrior who rises to the occasion. You'll find gratification in the journey—the path through adversity. It is my belief that the spirit of humanity thrives on the challenge of beating the odds. We all love a good underdog story, the hero's journey. Life would be boring if it was easy.

There's no finish line to this fight. This battle to guard your mind against Big Tech is an ongoing process, a constant cycle. Life has its ebbs and flows, points at which the struggle will be easier or more difficult. Use all of the tools at your disposal. Refine these or find new ones and share with the community.

No one is impervious to mental manipulation. Everyone is impacted at some level. I know so many people who spend hours a week training and maintaining their physical health but are unaware of what can atrophy and the vulnerabilities that come from living in a digital age.

Don't neglect the thing that's the essence of your existence: your mind.

ACKNOWLEDGMENTS

To my first and Top 8 friend of all time, **Tom Anderson**, thank you. You inspired me, along with an entire generation of developers and hackers, to build and create.

During my research I found so much meaningful insight from others. I am grateful to be able to reference, cite, and build on the exceptional work you have done: **Anna Lembke, MD**, author of *Dopamine Nation*; **Jonathan Haidt, MD**, author of *The Anxious Generation* and *The Coddling of the American Mind*; and **Cal Newport**, author of *Digital Minimalism* and *Slow Productivity*.

To **Andrew Huberman** and the *Huberman Lab* podcast, for your exceptional work. You continue to inspire me and others to learn, understand, and improve.

To my **mom**, for your love, unwavering support, and every sacrifice over the years. Thank you for being my biggest fan.

To my **sisters**, I love you.

To my brother-in-law **Chris**, for our conversations during this process and every bit of help along the way.

To my **Aunt Lorraine**, who got me my first computer. You helped open a door, and I am forever grateful.

To **Sharon Huey**, for making up elective classes where I was literally the only student so that I could pursue filmmaking. You dumped a tanker full of gas on my spark of creativity. Thank you.

To **Darin** and **Chris Rogers**, for the tenderloin biscuits, long nights chatting it up at the Raceway gas station, and the even later-night trips to the Awful Waffle.

To **Jeffrey Marcus**, for introducing me to the world of meditation and breath work and for broadening my capacity for love through perspectives on life and people.

To **Michael Gallagher**, for taking a risk on a goofy no-name comedian and giving me an opportunity and a platform to vent creatively.

To **Dave Eaton**, for all your hard work on so many videos and being one of the best friends I've ever known and the kindest and genuine souls I've ever met.

To **Matt Labate**, **Jim Louderback**, and **Gabe Magana** at Discovery Digital Networks, for believing in a creator so much that you were willing to loan me money against my future ad revenue so I could invest in myself and what we were doing.

To **Shana McInnes** and **Brandon Robinson**, for your tireless efforts and exceptional skills in everything we've done over so many years at so many companies.

To **Phil** and **Lindsay DeFranco**, for sharing your home with this hillbilly and his mutts while we embarked on this journey. Thank you for your love and friendship.

To **Neeraj Khemlani**, for your trust and support in me as an executive and my ability to perform at the highest level.

To **Mat Best**, **Evan Hafer**, **Logan Stark**, and **Jarred Taylor**, for the wild ride that was Black Rifle Coffee.

To **Jack Carr**, for your kind words of wisdom for me in this process and the example you set by the prolific work you do.

To **Jesse Griffiths**, for the food, the laughs, and the culinary and butchering skills.

To **Mike Ritland**, for Team Dog, the *Mike Drop* podcast, and everything you do for retired working dogs.

To **Robert Bigando**, for your enthusiasm and expertise on things that go boom.

To **Daniel Holloway**, for every conversation we get to have on and off the *Citizen Podcast*.

To **Ryan Morris**, for a place to sleep whenever I want to stay in a haunted bunkhouse and magazines filled with as much RDX and det cord as I could ever need.

To **Matthew "Woody" Woodworth, Kyle "FPSRussia" Myers**, **Taylor "Murka" Smallwood**, and **Christian "Chiz" Sirpilla**, for having me on the *PKA* podcast over the years and the assist on quotes.

To **Kevin Espiritu** and the entire Epic Gardening and Botanical Interests team, for your passion, vast knowledge, and help getting the perfect gifts to go with this project.

To **Anthony Pompliano**, for being a signal in a world of noise and bringing me on your podcast to talk about these things that matter to us both.

To **Brandon Herrera**, for walking the walk, carrying the torch, and standing up for what you believe in during a time when others are only willing to talk.

To **Cody Garrett**, for the wellness checks and invites to do fun things that occasionally force me out of my cave.

ACKNOWLEDGMENTS

To **DJ Shipley** and **Cole Fackler**, for the skydiving knowledge transfer and everything you do with GBRS.

To **Tim Kennedy**, for always bringing the energy and the best version of yourself every time we have worked together, from Ultimate Soldier Challenge to Veterans React.

To **Jamey Shirah**, for the absurdly large case of apple fritters to help me through this process with the book.

To **Baker Leavitt**, for just being you. For being brutally honest, helping coordinate efforts on this, and being a great friend.

To **Ben Levy** and **Brandon Kuipers**, for nudging me down this path with my book and for our pickup basketball sessions.

To **Sam Parr**, for coming in clutch at the midnight hour. There are many things I can thank you for, but mostly, thanks for being my friend.

To **Murphy Wilt**, for your hard work in all things, but especially in this endeavor.

To **Rachael Williams**, for your ironclad patience as my publishing manager.

To **Gail Fay**, for everything you have done for me on this journey as my scribe and psychiatrist. I am forever grateful.

To **Oswald** and **Kiwi,** for the stark reminder of the short window of time we get to share together on this Earth.

ABOUT THE AUTHOR

Richard Ryan is a software developer and media executive with more than fifteen years of experience in the tech industry. He has generated billions of views and millions of followers across social media platforms, leveraging his deep understanding of algorithms and digital marketing. As a co-founder of Black Rifle Coffee Company, Richard played a pivotal role in growing it into a publicly traded entity. His unique perspective stems from his expertise in subjects such as programming and executive leadership, making him a credible voice in the intersection of technology and behavior. Richard dedicates his efforts toward empowering individuals to become conscious consumers of digital content.

THIRTY-DAY CHALLENGE

As you read in Chapter 5, four weeks seems to be the optimum time for a dopamine reset. So I'm including a Thirty-Day Challenge template to help you with this detox process. Here you can track your usage insights, reflect on how detox is going, work on gratitude, complete your exercises, and respond to the journaling prompts. I still think you should buy a real journal, but the following pages provide a good starting point if you're new to the practice.

YESTERDAY'S INSIGHTS

Use your phone to fill out the Yesterday's Insights section:

- **Screen Time**: the phone setting that runs in the background and monitors your usage. Toggle between the Daily and Weekly tabs to see your usage by time and app.
  - iOS: Settings > Screen Time > See All App & Website Activity
  - Android: Settings > Digital Wellbeing & Parental Controls

- **Pickups** (iOS) or **Unlocks** (Android): the Screen Time statistic that monitors how often you wake your phone; in iOS, it also records the apps you launch immediately upon doing so. By tracking this metric, you gain insight on your potential subconscious and/or compulsive behavior around the device. This can also provide insight as to which applications are consuming most of your attention along with those you might have a potential addiction around.

- **Notifications** (IOS and Android): the Screen Time statistic that monitors how many notifications you are receiving from various applications on your device. These vary from visual to audible cues that can interrupt your thought processes by context switching throughout your day that can lead to a lack of productivity and or mental fatigue.

Tracking these metrics will give you insight into how your device is consuming your attention, your time, and your life. Then you can make decisions about what to change.

GRATITUDE

Use the gratitude prompts to start working on a daily gratitude practice. Be as specific or generic as you want.

There is no right or wrong way to do them. By practicing thankfulness, you will subconsciously program yourself to take note throughout your day of things to be grateful for. As shown in Chapter 9, studies find this can increase a person's happiness and overall well-being.

JOURNAL

On the second page of each day's template, I've included a Journal page. On some days, I've included one of the exercises or prompts that appear in each chapter (with the chapter number so you can look back for a more detailed explanation). On other days, I've left the Journal page completely blank.

No matter what the wording on any particular day says, you can write whatever you want. There's no right or wrong way to journal. You can respond to the exercise or prompt, or write about your feelings, your day, your thoughts about this whole assessment/detox process, or whatever is on your mind. The goal is to be more in tune with your thoughts, feelings, values, and priorities by day thirty so you can more effectively assess and address your relationship with your devices and digital media.

DAY
01
_5_ / _22_ / 20_10_
S M T W T (F) S

YESTERDAY'S INSIGHTS

Total Numbers

Screen Time: 5 hours 32 minutes

Pickups/Unlocks: 228

Notifications: 182

Top Three Apps

Most Used	First Used	Notifications
1. Messages	1. Messages	1. Messages
2. Pose Ai	2. Instagram	2. Amazon
3. TikTok	3. Calendar	3. Door Dash

GRATITUDE

Who are you grateful for in your life?

My health. I often forget that until I'm injured or sick. I'm also really grateful for my family and that I get to spend another day with them.

TODAY I FEEL

JOURNAL

Prompt 1: Know Yourself. Answer these questions to prime the self-reflection pump (Chapter 3).

....
....
....
....
....
....
....
....
....
....
....
....
....
....
....
....
....

DAY
01 ___/___/20___
S M T W T F S

YESTERDAY'S INSIGHTS
Total Numbers

Screen Time: ....

Pickups/Unlocks: ....

Notifications: ....

Top Three Apps

Most Used	First Used	Notifications
1.	1.	1. ....
2.	2.	2. ....
3.	3.	3. ....

GRATITUDE

Who are you grateful for in your life?

....

....

....

....

TODAY I FEEL

JOURNAL

Prompt 1: Know Yourself. Answer these questions to prime
the self-reflection pump (Chapter 3).

....

....

....

....

....

....

....

....

....

....

....

....

....

....

....

....

....

....

....

DAY 02

___/___/20___

S M T W T F S

YESTERDAY'S INSIGHTS

Total Numbers

Screen Time: ....

Pickups/Unlocks: ....

Notifications: ....

Top Three Apps

Most Used	First Used	Notifications
1.	1.	1. ....
2.	2.	2. ....
3.	3.	3. ....

GRATITUDE

What are you grateful for in your life?

....

....

....

....

TODAY I FEEL

JOURNAL

Exercise 1: Are You Addicted? Assess whether you subconsciously and/or compulsively rely on your phone (Chapter 3).

....

....

....

....

....

....

....

....

....

....

....

....

....

....

....

....

....

DAY
03
___/___/20___
S M T W T F S

YESTERDAY'S INSIGHTS

Total Numbers

Screen Time: ....

Pickups/Unlocks: ....

Notifications: ....

Top Three Apps

Most Used	First Used	Notifications
1.	1.	1. ....
2.	2.	2. ....
3.	3.	3. ....

GRATITUDE

Who are you grateful for today?

....

....

....

....

TODAY I FEEL

😣 😟 😐 🙂 😃

JOURNAL

Prompt 2: Screen Time Reality Check. Use your Screen Time numbers to reflect on the consequences of your time usage and how you might use that time differently (Chapter 3).

....

....

....

....

....

....

....

....

....

....

....

....

....

....

....

....

....

DAY
04 ___/___/20___
S M T W T F S

YESTERDAY'S INSIGHTS
Total Numbers

Screen Time: ....

Pickups/Unlocks: ....

Notifications: ....

Top Three Apps

Most Used	First Used	Notifications
1.	1.	1. ....
2.	2.	2. ....
3.	3.	3. ....

GRATITUDE
What are you grateful for today?

....

....

....

....

TODAY I FEEL

JOURNAL

....

....

....

....

....

....

....

....

....

....

....

....

....

....

....

....

....

....

....

....

DAY
05

___/___/20___

S M T W T F S

YESTERDAY'S INSIGHTS

Total Numbers

Screen Time: ....

Pickups/Unlocks: ....

Notifications: ....

Top Three Apps

Most Used	First Used	Notifications
1.	1.	1. ....
2.	2.	2. ....
3.	3.	3. ....

GRATITUDE

Who would you be sad about losing if they were no longer in your life?

....

....

....

....

TODAY I FEEL

JOURNAL

Exercise 2: Values and Priorities. Write out your core values and priorities (Chapter 4).

....

....

....

....

....

....

....

....

....

....

....

....

....

....

....

....

....

....

DAY 06

___/___/20___

S M T W T F S

YESTERDAY'S INSIGHTS

Total Numbers

Screen Time: ....

Pickups/Unlocks: ....

Notifications: ....

Top Three Apps

Most Used   First Used   Notifications

1. ....   1. ....   1. ....

2. ....   2. ....   2. ....

3. ....   3. ....   3. ....

GRATITUDE

What would you be sad about losing if it were no longer in your life?

....

....

....

....

TODAY I FEEL

JOURNAL

Prompt 3: Letters to Yourself (part 1). Write a letter to fifteen-year-old you (Chapter 4).

....

....

....

....

....

....

....

....

....

....

....

....

....

....

....

....

....

....

DAY
07 __/__/20__
S M T W T F S

YESTERDAY'S INSIGHTS
Total Numbers

Screen Time: ....

Pickups/Unlocks: ....

Notifications: ....

Top Three Apps

Most Used	First Used	Notifications
1.	1.	1. ....
2.	2.	2. ....
3.	3.	3. ....

GRATITUDE
What made you smile today?

....

....

....

....

TODAY I FEEL

JOURNAL

Prompt 3: Letters to Yourself (part 2). Write a letter to ninety-two-year-old you (Chapter 4).

DAY
08 ___/___/20___
S M T W T F S

YESTERDAY'S INSIGHTS
Total Numbers

Screen Time: ....

Pickups/Unlocks: ....

Notifications: ....

Top Three Apps

Most Used	First Used	Notifications
1.	1.	1. ....
2.	2.	2. ....
3.	3.	3. ....

GRATITUDE
Who are you grateful for in your life?

....

....

....

....

TODAY I FEEL

JOURNAL

....

....

....

....

....

....

....

....

....

....

....

....

....

....

....

....

....

....

....

....

....

....

DAY
09

___/___/20___

S M T W T F S

YESTERDAY'S INSIGHTS
Total Numbers

Screen Time: ....

Pickups/Unlocks: ....

Notifications: ....

Top Three Apps

Most Used	First Used	Notifications
1.	1.	1. ....
2.	2.	2. ....
3.	3.	3. ....

GRATITUDE
What are you grateful for in your life?

....

....

....

....

TODAY I FEEL

JOURNAL

Exercise 3: What and How. Decide what aspects of your digital life you need to detox from and decide how you're going to do that (Chapter 5).

....

....

....

....

....

....

....

....

....

....

....

....

....

....

....

....

....

....

....

DAY
10 ___/___/20___
S M T W T F S

YESTERDAY'S INSIGHTS
Total Numbers

Screen Time: ....

Pickups/Unlocks: ....

Notifications: ....

Top Three Apps

Most Used	First Used	Notifications
1.	1.	1. ....
2.	2.	2. ....
3.	3.	3. ....

GRATITUDE
Who are you grateful for in your life?

....

....

....

....

TODAY I FEEL

JOURNAL

Prompt 4: Doing Hard Things. Reflect on the last time you did something truly difficult—why it was so hard, how long it took you, and how you felt when you finished (Chapter 5).

....

....

....

....

....

....

....

....

....

....

....

....

....

....

....

....

....

DAY
11
___/___/20___
S M T W T F S

YESTERDAY'S INSIGHTS
Total Numbers

Screen Time: ....

Pickups/Unlocks: ....

Notifications: ....

Top Three Apps

Most Used	First Used	Notifications
1.	1.	1. ....
2.	2.	2. ....
3.	3.	3. ....

GRATITUDE
What are you grateful for today?

....

....

....

....

TODAY I FEEL

JOURNAL

DAY
12 ___/___/20___
S M T W T F S

YESTERDAY'S INSIGHTS

Total Numbers

Screen Time: ....

Pickups/Unlocks: ....

Notifications: ....

Top Three Apps

Most Used	First Used	Notifications
1.	1.	1. ....
2.	2.	2. ....
3.	3.	3. ....

GRATITUDE

Who would you be sad about losing if they were no
longer in your life?

....

....

....

....

TODAY I FEEL

JOURNAL

Exercise 4: Pick Three. Pick three specific friction actions you're going to start using right now (Chapter 6).

....

....

....

....

....

....

....

....

....

....

....

....

....

....

....

....

....

DAY
13 ___/___/20___
S M T W T F S

YESTERDAY'S INSIGHTS
Total Numbers

Screen Time: ....

Pickups/Unlocks: ....

Notifications: ....

Top Three Apps

Most Used	First Used	Notifications
1.	1.	1. ....
2.	2.	2. ....
3.	3.	3. ....

GRATITUDE
What would you be sad about losing if it were no longer in your life?

....

....

....

....

TODAY I FEEL

JOURNAL

Prompt 5: Make It Personal. Think about why you need friction in the areas you picked, what the result will be, and why taking this step is important (Chapter 6).

....

....

....

....

....

....

....

....

....

....

....

....

....

....

....

....

....

....

DAY
14 ___/___/20___
S M T W T F S

YESTERDAY'S INSIGHTS

Total Numbers

Screen Time: ....

Pickups/Unlocks: ....

Notifications: ....

Top Three Apps

Most Used	First Used	Notifications
1.	1.	1. ....
2.	2.	2. ....
3.	3.	3. ....

GRATITUDE

What do you love about where you live?

....

....

....

....

TODAY I FEEL

😣 🙁 😐 🙂 😄

JOURNAL

DAY
15 ___ / ___ / 20 ___
S M T W T F S

YESTERDAY'S INSIGHTS
Total Numbers

Screen Time: ....

Pickups/Unlocks: ....

Notifications: ....

Top Three Apps

Most Used	First Used	Notifications
1.	1.	1. ....
2.	2.	2. ....
3.	3.	3. ....

GRATITUDE
Who are you grateful for in your life?

....

....

....

....

TODAY I FEEL

JOURNAL

Exercise 5: Potential Partners. List three potential account-ability partners and give yourself a deadline for picking one and approaching them (Chapter 7).

....

....

....

....

....

....

....

....

....

....

....

....

....

....

....

....

....

DAY 16

___/___/20___

S M T W T F S

YESTERDAY'S INSIGHTS

Total Numbers

Screen Time: ....

Pickups/Unlocks: ....

Notifications: ....

Top Three Apps

Most Used	First Used	Notifications
1.	1.	1. ....
2.	2.	2. ....
3.	3.	3. ....

GRATITUDE

What are you grateful for in your life?

....

....

....

....

TODAY I FEEL

JOURNAL

Prompt 6: Create an Accountability Plan. Decide how often you'll meet with your accountability partner, what you'll be held accountable for, how you'll provide proof, and how you feel about this whole accountability thing (Chapter 7).

DAY
17
___/___/20___
S M T W T F S

YESTERDAY'S INSIGHTS

Total Numbers

Screen Time: ....

Pickups/Unlocks: ....

Notifications: ....

Top Three Apps

Most Used	First Used	Notifications
1.	1.	1. ....
2.	2.	2. ....
3.	3.	3. ....

GRATITUDE

Who are you grateful for today?

....

....

....

....

TODAY I FEEL

😣 🙁 😐 🙂 😃

JOURNAL

DAY
18

___/___/20___

S M T W T F S

YESTERDAY'S INSIGHTS

Total Numbers

Screen Time: ....

Pickups/Unlocks: ....

Notifications: ....

Top Three Apps

Most Used   First Used   Notifications

1. ....   1. ....   1. ....

2. ....   2. ....   2. ....

3. ....   3. ....   3. ....

GRATITUDE

What are you grateful for today?

....

....

....

....

TODAY I FEEL

JOURNAL

Exercise 6: Find Your Tribe. List five activities, hobbies, clubs, or groups you're going to explore and when you're going to get involved (Chapter 8).

....

....

....

....

....

....

....

....

....

....

....

....

....

....

....

....

....

....

DAY
19

___/___/20___

S M T W T F S

YESTERDAY'S INSIGHTS

Total Numbers

Screen Time: ....

Pickups/Unlocks: ....

Notifications: ....

Top Three Apps

Most Used	First Used	Notifications
1.	1.	1. ....
2.	2.	2. ....
3.	3.	3. ....

GRATITUDE

Who would you be sad about losing if they were no longer in your life?

....

....

....

....

TODAY I FEEL

JOURNAL

Prompt 7: Staying Balanced. Write about a time when you felt part of a community: who you were, what your life was like, why this moment was significant, and how you felt (Chapter 8).

....

....

....

....

....

....

....

....

....

....

....

....

....

....

....

....

DAY 20

___/___/20___

S M T W T F S

YESTERDAY'S INSIGHTS

Total Numbers

Screen Time: ....

Pickups/Unlocks: ....

Notifications: ....

Top Three Apps

Most Used	First Used	Notifications
1.	1.	1. ....
2.	2.	2. ....
3.	3.	3. ....

GRATITUDE

What would you be sad about losing if it were no longer in your life?

....

....

....

....

TODAY I FEEL

JOURNAL

....

....

....

....

....

....

....

....

....

....

....

....

....

....

....

....

....

....

....

....

DAY 21

___ / ___ / 20___

S M T W T F S

YESTERDAY'S INSIGHTS

Total Numbers

Screen Time: ....

Pickups/Unlocks: ....

Notifications: ....

Top Three Apps

Most Used	First Used	Notifications
1.	1.	1. ....
2.	2.	2. ....
3.	3.	3. ....

GRATITUDE

What opportunities have you had that you are
grateful for?

....

....

....

....

TODAY I FEEL

JOURNAL

Exercise 7: Start Practicing. List three ways you will incorporate gratitude into your daily routine and state when you will start (Chapter 9).

....

....

....

....

....

....

....

....

....

....

....

....

....

....

....

....

....

DAY
22

___/___/20___

S M T W T F S

YESTERDAY'S INSIGHTS

Total Numbers

Screen Time: ....

Pickups/Unlocks: ....

Notifications: ....

Top Three Apps

Most Used	First Used	Notifications
1.	1.	1. ....
2.	2.	2. ....
3.	3.	3. ....

GRATITUDE

Who are you grateful for in your life?

....

....

....

....

TODAY I FEEL

JOURNAL

Prompt 8: Gratitude for the Little Things. Write about a time when someone helped you and how you felt in the moment (Chapter 9).

....

....

....

....

....

....

....

....

....

....

....

....

....

....

....

....

....

....

DAY
23 ___/___/20___
S M T W T F S

YESTERDAY'S INSIGHTS
Total Numbers

Screen Time: ....

Pickups/Unlocks:..

Notifications: ....

Top Three Apps

Most Used   First Used   Notifications

1. ....   1. ....   1. ....

2. ....   2. ....   2. ....

3. ....   3. ....   3. ....

GRATITUDE
What are you grateful for in your life?

....

....

....

....

TODAY I FEEL

JOURNAL

DAY 24

___/___/20___
S M T W T F S

YESTERDAY'S INSIGHTS

Total Numbers

Screen Time: ....

Pickups/Unlocks: ....

Notifications: ....

Top Three Apps

Most Used	First Used	Notifications
1.	1.	1. ....
2.	2.	2. ....
3.	3.	3. ....

GRATITUDE

Who are you grateful for today?

....

....

....

....

TODAY I FEEL

JOURNAL

Prompt 7: Staying Balanced. Write about another time when you felt part of a community: who you were, what your life was like, why this moment was significant, and how you felt (Chapter 8).

....

....

....

....

....

....

....

....

....

....

....

....

....

....

....

....

DAY
25
___/___/20___

S M T W T F S

YESTERDAY'S INSIGHTS

Total Numbers

Screen Time: ....

Pickups/Unlocks: ....

Notifications: ....

Top Three Apps

Most Used	First Used	Notifications
1.	1.	1. ....
2.	2.	2. ....
3.	3.	3. ....

GRATITUDE

What are you grateful for today?

....

....

....

....

TODAY I FEEL

JOURNAL

DAY
26 ___/___/20___
S M T W T F S

YESTERDAY'S INSIGHTS
Total Numbers

Screen Time: ....

Pickups/Unlocks: ....

Notifications: ....

Top Three Apps

Most Used	First Used	Notifications
1.	1.	1. ....
2.	2.	2. ....
3.	3.	3. ....

GRATITUDE
Who would you be sad about losing if they were no longer in your life?

....

....

....

....

TODAY I FEEL

JOURNAL

Prompt 8: Gratitude for the Little Things. Write about another time when someone helped you and how you felt in the moment (Chapter 9).

DAY
27
___/___/20___
S M T W T F S

YESTERDAY'S INSIGHTS

Total Numbers

Screen Time: ....

Pickups/Unlocks: ....

Notifications: ....

Top Three Apps

Most Used	First Used	Notifications
1.	1.	1. ....
2.	2.	2. ....
3.	3.	3. ....

GRATITUDE

What would you be sad about losing if it were no longer in your life?

....

....

....

....

TODAY I FEEL

JOURNAL

....

....

....

....

....

....

....

....

....

....

....

....

....

....

....

....

....

....

....

....

DAY 28

___/___/20___

S M T W T F S

YESTERDAY'S INSIGHTS

Total Numbers

Screen Time: ....

Pickups/Unlocks: ....

Notifications: ....

Top Three Apps

Most Used	First Used	Notifications
1.	1.	1. ....
2.	2.	2. ....
3.	3.	3. ....

GRATITUDE

What life lessons are you grateful for?

....

....

....

....

TODAY I FEEL

JOURNAL

DAY
29
___/___/20___
S M T W T F S

YESTERDAY'S INSIGHTS

Total Numbers

Screen Time: ....

Pickups/Unlocks: ....

Notifications: ....

Top Three Apps

Most Used	First Used	Notifications
1.	1.	1. ....
2.	2.	2. ....
3.	3.	3. ....

GRATITUDE

Who are you grateful for in your life?

....

....

....

....

TODAY I FEEL

JOURNAL

....

....

....

....

....

....

....

....

....

....

....

....

....

....

....

....

....

....

....

....

DAY
30 ___/___/20___
S M T W T F S

YESTERDAY'S INSIGHTS
Total Numbers

Screen Time: ....

Pickups/Unlocks: ....

Notifications: ....

Top Three Apps

Most Used	First Used	Notifications
1.	1.	1. ....
2.	2.	2. ....
3.	3.	3. ....

GRATITUDE
What are you grateful for in your life?

....

....

....

....

TODAY I FEEL

JOURNAL